Enforcing Covenants: A Guide for Managers of Leasehold Property

Brian Jones

2005

A division of Reed Business Information

Estates Gazette
1 Procter Street, London WC1V 6EU

ISBN 0 7282 0483 5

Typeset in Palatino 10/12 by Amy Boyle, Rochester
Printed by Bell & Bain Ltd., Glasgow

Contents

Acknowledgements

I would like to thank the following for all their help towards the writing and publication of this book.

Alison, Audrey and Amy and everyone associated with Estates Gazette Books.

My associates, John Nouch, Andrew McKeer and Jeff Platt, for their support and encouragement.

Neil Maloney of Granvilles and Brett Williams and Karen Combellack of Curry and Partners for suggesting some case studies.

David Sanderson and Marie Batchelor of Shoosmiths for providing some draft precedents.

The staff of the Office of Public Sector Information, HM Courts Service and the Residential Property Tribunals Service for their guidance on reproduction of court and LVT forms and explaining Crown copyright to me.

My family for all their support and patience, and help with the odd bit of research here and there.

And especially my wife, District Judge Sally Levinson, for all her help, including putting up with my interminable questions checking on current county court practice. However I take sole responsibility for the contents of this book. Apart from direct quotations, the text is mine and mine alone.

Table of Statutes

Table of Statutory Instruments

Table of Cases

Leasehold Valuation Tribunal decisions

Foreword

Living in a block of flats is in someways like living in a microcosm of society as a whole; you have a defined community that will only exist in peace and harmony and in economic equilibrium if all those members of that community abide by the "rules".

Sadly in society today there are those do who not abide by such rules and so laws and sanctions have to be introduced, which often unintentionally also affect those who do conform.

In blocks of flats where the residents "own" their units on a long lease the situation is compounded by the fact that *their* homes and communities are, more often than not, managed by others, whether an investor freeholder or a residents' management company, which in turn utilises the services of a managing agent.

In the case of this book the "rules" are the covenants in the lease which is a contract setting out the rights and obligations of the lessor and lessee; and if well drafted the covenants will provide a successful framework for interdependent living — but only if both parties comply with them. Regrettably this is not always the case and therefore a raft of legislation has come into being to overlay the provisions in the lease.

Covenants have to be enforced or a block of leasehold flats becomes unmanageable with the disastrous result that the value of the flats can be materially affected and the quality of life of the residents deteriorates. In this enlightened day and age of residential leasehold management it is vital that enforcement is done "efficiently so, it can at least be tolerable and cause minimal damage to relationships between landlord, manager and leaseholder".

These are words I have taken from Brian Jones' preface of this, his second excellent book dealing with residential leasehold issues. Brian is well known in our sector generally and well respected by the Association of Residential Managing Agents (ARMA) and its members as a lawyer not only thoroughly versed in the hard facts of legislation but also the purely practical issues of good day-to-day management.

Brian's approach in *Enforcing Covenants: A Property Manager's Guide* is based on the law as it now stands but takes the approach of how this law, and the covenants in the lease, can be enforced in the most efficient manner but with least damage to all the parties concerned.

The book contains case studies to help the reader visualise how enforcement works in practice and, of course, is sprinkled with Brian's own kind of humour that is certainly appreciated by the very many audiences that have heard him speak.

I cannot say there is any book on residential leasehold that I would recommend as bedtime reading, but Brian's style and approach to this difficult area of property management makes it a good and worthwhile read!

David C Hewett FIRPM
Executive Secretary, ARMA

Preface

On 28 February 2005, a major portion of the Commonhold and Leasehold Reform Act 2002 came into force, including those sections which bring about fundamental changes to ground rent and forfeiture for landlords, leaseholders and managers of residential leasehold properties. These changes were flagged up in Chapters 15 and 16 of my previous book: *Right to Manage and Service Charges: The New Regime.* In very general terms, the collection and recovery of ground rents will become more unwieldy and enforcement of leaseholders' covenants will be more complicated, more expensive and less sure of success.

Potentially, the effects could be worst upon leaseholders managing their own properties, whether through residents' management companies, right to manage companies or the new right to enfranchise companies.

The purpose of *Enforcing Covenants* is to analyse in greater depth the meaning and effects of these specific changes, to extrapolate the effects into existing law and practice in this field and then to develop practical strategies in response for those involved in the management of blocks of flats. It should be of use to landlords, managing agents, directors of residents' management companies, individual leaseholders, right to manage companies, right to enfranchise companies and the professional and legal advisers of all these parties.

In addition to the text, I have gathered together a selection of case studies, precedent forms and documents which are intended as a guide to action rather than as recipes for precise reproduction in each case. The characters, names, events and situations portrayed in the case studies are entirely fictitious and imaginary. Readers who have had the pleasure of hearing my contributions at conferences and seminars over the years may recognise some of the names — indeed, Tumbledown Mansions and its inhabitants are becoming worryingly real to me — I can confirm however that any similarities between any of the characters or events in this book and any real persons or incidents is entirely coincidental.

I have not attempted to cover in any depth the question of enforcement as against landlords, partly because of the simple matter of space and partly because the changes in the law have had a far greater impact upon enforcement against leaseholders. In any event, the subject is covered in many ways by my previous book *Right to Manage and Service Charges: The New Regime.*

Enforcement is an unavoidable task for anyone involved actively in management of leasehold property. It is never likely to be pleasant. If it is done efficiently however, it can at least be tolerable and cause minimal damage to relationships between landlord, manager and leaseholder. I hope that this book will help to enhance the effectiveness of those who manage, and advance the understanding of what is involved for other interested parties.

Brian Jones
August 2005

Covenants 1

Before considering how covenants may be enforced, it is necessary to establish how they come about.

What is a covenant?

Put very simply, a lease is a contract by which a landlord grants to a purchaser a right to use and occupy part of his land for a fixed term of years in return for some form of consideration (in a long lease the consideration will normally comprise the payment of a premium and the agreement to pay rent). The "landlord" or "lessor" may be the freeholder of the land, or he may hold a superior lease (a "headlease"). The purchaser may be described as "leaseholder", "lessee" or "tenant".

With a view to ensuring the smooth running of the contract over the term of the lease (which is likely to be at least 99 years) the lease contains an exchange of promises by each party. These contractual promises (or "covenants") will be binding not only upon the original parties but also on any subsequent assignees, purchasers or successors of their respective interests. The parties may covenant to perform specific acts or not to do certain things or even not to permit others to act in a particular way.

If a covenant is broken (or "breached"), its performance can be enforced by the innocent party. Indeed, if the breach is sufficiently serious (but only of a leaseholder's covenant), the landlord can bring about an early termination of the lease by way of forfeiture; however, the availability of this ultimate sanction is very severely limited, as we shall see.

To be enforceable, the wording of the covenant must be sufficiently clear and unambiguous; if its meaning cannot be ascertained it cannot be enforced. Therefore, the party seeking to rely upon a covenant must establish that it is in the lease and that their interpretation of its meaning and effect stands up to scrutiny by the courts or tribunals. (This may seem obvious, but it is not uncommon for landlords or tenants to attempt to sue upon a complete misreading or misunderstanding of the terms of their lease.)

Occasionally, Acts of Parliament or judicial decisions will provide that particular covenants will be implied into leases, whether they are expressly contained in the lease or not. These "implied covenants" tend to relate to repairing issues (such as under section 4 of the Defective Premises Act 1972) or the way premises are used.

What different types of covenant are typically found in long residential leases?

A: By the landlord:

- to maintain, repair and redecorate the main structure and common parts of the building
- to insure the building
- to collect and account for service charges
- to give the tenant "quiet enjoyment" of the demised property
- to enforce covenants against other tenants of the landlord (at the tenant's cost)

B: By the tenant:

- to pay rent
- to pay service charges
- to pay rates, taxes and similar charges imposed upon the demised property by central and local government
- not to make any structural alteration
- to pay costs relating to notices under Section 146 of the Law of Property Act 1925 (see below)
- to give the landlord notice of any dealings in the property
- to repair and decorate internally
- to permit access to the landlord for repairs or inspection
- not to do anything which might render buildings insurance void or voidable

- not to cause nuisance or annoyance to other residents
- to observe regulations (generally by reference to a schedule).

As can be seen, the tenant's covenants massively outnumber the landlord's.

C: By a management company (where it is a party to the lease):

- to maintain, repair and redecorate the main structure and common parts of the building
- to insure the building
- to collect and account for service charges.

What powers of enforcement are typically found in a lease?

It is extremely unusual to find anything regarding the landlord's covenants with the possible exception of an arbitration clause relating to disputes between landlord and tenant. Arbitration clauses are looked at in chapter 12. As far as management companies are concerned, there may be provision for the landlord to step in if they fail.

It will be an extremely unusual lease however which does not include a proviso for re-entry by the landlord in the event of breach of the tenant's covenants ("the forfeiture clause"). Forfeiture will be examined in more depth in the next chapter.

Generally, there will be a covenant by the tenant to pay the landlord's costs and expenses relating to forfeiture or incurred in anticipation of such an event. The word "forfeiture" may not be used, but that is the general intent. There is a small variety of typical costs clauses. Common examples include:

- costs incurred "incidental to the preparation and service of a notice under section 146 of the Law of Property Act 1925"
- "costs of and in contemplation of proceedings under section 146"
- costs relating to "proceedings for the recovery of the rents"
- costs arising "by reason of any breach of any of the tenant's covenants notwithstanding that forfeiture may be avoided otherwise than by relief granted by the court".

Some leases even contain more than one such clause. As will be seen later, their effectiveness has been eroded by legislation and by judgments of the courts over the last few years.

Many leases provide that the landlord may sue for reimbursement of expenditure in carrying out the tenant's repairing responsibilities in the event that the tenant fails to do so. Such clauses tend to reserve the landlord's right to sue for a debt in such a case whereas the normal remedy for breach of a covenant in a contract is damages.

Finally, leases will frequently contain penalty clauses for late payment of rents and service charges. Most typically these allow landlords to charge interest on arrears at, say, 4% above base rate. Although the Commonhold and Leasehold Reform Act 2002 generally frowns upon penalty charges, section 162(4) recognises that an argument can be made out that interest should be permitted by adding it to the grounds in section 35 of the Landlord and Tenant Act 1987 by which a lease may be varied.

Forfeiture

What is forfeiture?

Forfeiture is a power in favour of the landlord to bring the lease to an early end and regain possession of the premises without compensation to the lessee or anyone deriving title under him (such as a mortgage lender or subtenant). As a weapon, forfeiture (or indeed merely the threat of it) is formidable; however the power arises only if certain circumstances exist and it is subject to an array of exceptions, conditions and restrictions imposed by law.

Who can forfeit?

It is important to remember that the right of forfeiture is vested only in the party with freehold title (or the immediate landlord if an under-lease) no matter who is entitled directly to the benefit of the lessee's covenants. For example, a right to manage ("RTM") company acquires all the landlord's management functions and can enforce the lessee's covenants *except by forfeiture*. If an RTM company wishes to threaten forfeiture, it must have the landlord's express consent and co-operation. The same would apply to a manager appointed by the Leasehold Valuation Tribunal ("LVT") or to a Residents' Management Company ("RMC"), which is a party to the lease as manager only.

Of course, if a RMC or other company made up of leaseholders acquires the freehold title by enfranchisement or otherwise, it will also acquire the landlord's right of forfeiture. Unless and until that happens, any form of manager or management company will either have to obtain the landlord's agreement (and any forfeiture proceedings would have to be in the landlord's name) or use alternative enforcement measures.

When can it be done?

First, forfeiture is not an option unless there is a forfeiture clause (the "proviso for re-entry") in the lease concerned. This must be checked if forfeiture is contemplated as its inclusion is not automatic nor is it implied. Indeed, it is not unheard of for the clause to be omitted in a lease for a single flat in a block where the other leases contain it.

An example of a forfeiture clause would bc:

> PROVIDED ALWAYS and it is hereby agreed that if the rent hereby reserved or any part thereof shall be unpaid for 14 days after becoming payable (whether formally demanded or not) or if any covenants on the part of the Tenant herein contained shall not be performed or observed then and in any such case it shall be lawful for the Landlord at any time thereafter to re-enter upon the Demised Premises or any part thereof in the name of the whole and thereupon this demise shall absolutely determine but without prejudice to any right of action or remedy of the Landlord in respect of any antecedent breach of any of the Tenant's covenants or the conditions herein contained.

Assuming that the clause exists, it is necessary to confirm that it covers the appropriate eventuality. Most forfeiture clauses will operate if rent has been unpaid for a specified number of days (14, 21 or 28 are the most common periods) or in the event of any breach or non-observance of the lessee's covenants. Occasionally, this will be limited by a further proviso; most typically that forfeiture will not be available if the only ground is unpaid interim service charges.

The other point to consider regarding the form of words used is the reference to rent — specifically "whether formally demanded or not". For centuries landlords have been entitled to forfeit leases for unpaid rent even if the amounts involved were almost trivial and without having to give prior notification or even a reminder. This was enshrined in statute (section 210, Common Law Procedure Act 1852). Parliament has now intervened to put a stop to that in residential leases.

Section 166 of the Commonhold and Leasehold Reform Act 2002 provides (with effect from 28 February 2005) that the liability to pay rent does not arise unless and until the landlord has issued a notice in the prescribed form (as set out in the Landlord and Tenant (Notice of Rent)

(England) Regulations 2004 SI 2004 No 3096). This provision overrides any conflicting provision in a lease so that words such as "whether formally demanded or not" will cease to have any practical effect.

How is it done?

The procedures for forfeiture have changed somewhat as a result of the implementation of sections 167 to 171 of the Commonhold and Leasehold Reform Act 2002 (the 2002 Act) on 28 February 2005. There are variations depending on whether the breach asserted is non-payment of rent, non-payment of service charges, failure to repair or failure to perform other covenants. The different procedures will be looked at in detail in Chapters 5 (ground rent), 6 (service charges), 8 (repairs) and 9 (other breaches), but the fundamental steps are summarised below.

A. *For covenants other than payment of rent*

1. If the breach is not admitted by the leaseholder it must first be determined to have occurred by the court or the LVT).
2. The landlord may not serve a notice until 14 days after such determination and the time for bringing an appeal. After this the landlord may serve a notice under section 146 of the Law of Property Act 1925 (a "section 146 notice" — this is not required if the breach is non-payment of monies reserved by the lease as additional rent — see below).
3. Copies of the notice should be served upon any mortgage lenders or subtenants who are interested in the property.
4. If, a "reasonable time" after service of the section 146 notice, the breach remains unremedied, forfeiture may be effected by "peaceable re-entry" (but only if the property is entirely clear of any vestiges of residential occupation) or by service of a county court claim for possession.
5. If there has been no successful defence or relief from forfeiture has not been granted (see below) so that an order for possession is obtained the landlord will be entitled to request the bailiffs to deliver up possession of the property to him.
6. The fact that the lease has been terminated by forfeiture is reported to H M Land Registry and application made to close the title.

7. The landlord is free to grant a new lease to a third party or deal in the property otherwise as he sees fit. No compensation is payable to the previous leaseholder, so the landlord may receive a substantial windfall if the losses suffered because of the breaches and the costs and the proceedings are far outstripped by the proceeds of the property (which will usually be the case).

Note: there is a special procedure for repairing breaches required by the Leasehold Property (Repairs) Act 1938 (see Chapter 8).

B. For non-payment of rent

1. The right of forfeiture does not arise unless rent has been demanded in accordance with section 166 of the 2002 Act and more than £350 is outstanding or arrears exceed three years (Section 167 of the 2002 Act).
2. No section 146 notice is required, but it is usual to give simple written notice of the landlord's intention to forfeit to any mortgage lender or subtenant — indeed, failure to do so will render the landlord liable for criticism and possible penalties in costs.
3. Forfeiture is effected by peaceable re-entry (if it is lawful) or the service of a claim for possession.
4. Steps 5 to 7 under A above will apply.

Monies reserved as rent or additional rent

A fairly high proportion of leases include within the definition of "rent" the payment of monies such as service charges or just the leaseholders' insurance contributions (commonly known as "insurance rent"). Reserving such payments as rent in this way will make some difference to the way they are treated, but possibly not as much as was originally intended.

For the purposes of service charge legislation, the definition which counts is that to be found in Section 18 of the Landlord and Tenant Act 1985 (as amended):

> (1) In the following provisions of this Act "service charge" means an amount payable by a tenant of a dwelling as part of or in addition to the rent —
>
> (a) which is payable, directly or indirectly, for services, repairs,

maintenance, improvements or insurance or the landlord's costs of management, and

(b) the whole or part of which varies or may vary according to the relevant costs.

(2) The relevant costs are the costs or estimated costs incurred or to be incurred by or on behalf of the landlord, or a superior landlord, in connection with the matters for which the service charge is payable.

(3) For this purpose —

(a) "costs" includes overheads, and

(b) costs are relevant costs in relation to a service charge whether they are incurred, or to be incurred, in the period for which the service charge is payable or in an earlier or later period.

Consequently, however they may be described in a particular lease, any charges to the leaseholder which come within the section 18 definition will be treated by the courts or tribunals in the same way as any other service charge when it comes to issues under their jurisdictions flowing from the relevant statutes from 1985 onwards. This includes any matters under the 2002 Act.

The same does not apply to the distinction (between service charges reserved as rent and service charges not treated by the lease in that way) as it affects the much earlier section 146 of the Law of Property Act 1925. Section 146(11) provides that the section 146 regime does not bear upon non-payment of rent, and that is still the case. As a result, it remains necessary to bear this in mind when drafting a S.146 Notice, and not to include within such a notice any outstanding sums which are reserved by the lease as rent. To do so may be regarded as oppressive and lead to penalties (if only in costs); moreover, it may lead to the wrong procedure being followed and thus undermine the whole of the landlord's case.

Purchaser's liability

The question of how and why a leaseholder who has acquired a property and then apparently becomes liable for his predecessor's breaches of covenant is one which has been shrouded in mystification for property managers for as long as there have been blocks of flats. It is not the purpose of this book to explain the legal position in "chapter and verse" — that would take too long with convoluted expositions on such matters as covenants running with the land (which has extant case law dating back to 1583) and the doctrines of privity of contract and privity of estate, and the situation has become even more muddied as a result of the Landlord and Tenant (Covenants) Act 1995; but a brief outline is necessary to provide a framework.

In the event of a breach of covenant, the landlord is entitled to elect to determine the lease by forfeiture. The essential question is not who is liable for the breach or even who committed it, but who needs to be evicted from the premises to complete the forfeiture? Unless, very unusually, it is possible to effect forfeiture by taking physical possession through peaceable re-entry, this really means who is the appropriate defendant in forfeiture proceedings? Obviously, it can only be the person currently entitled to possession. A forfeiture action is therefore effectively a claim against the land rather than the individual leaseholder, whoever that may be from time to time.

When it comes to exploring any remedy other than forfeiture, the position is somewhat different. A claim for a debt for example is made against the person rather than the land, so a leaseholder who has bought into a flat is not liable for his predecessor's debts as such. This is so even if the landlord attempts to levy distress for rent arrears (*Wharfland* v *South London Co-operative Building Co* [1995] 2 EGLR 21). The landlord could be in the position of having to trace the previous leaseholder to take proceedings against him (although the need to do this is extremely rare in practice). Having said all of that, if the breach of covenant concerned can be classed as a "continuing breach" then even if it originated under an earlier leaseholder the purchaser will be liable if it persists after the assignment.

Determining whether a breach is "continuing" or "once and for all" is not necessarily simple. For example, payment of service charges would appear to be an ongoing liability so non-payment could logically be regarded as continuing, enforceable at any time subject to limitation periods (see Chapter 6). On the other hand, failure to perform an act on or by a specified date would be a once and for all breach, and failure to pay a service charge on a specific quarter day could be seen in that light; in other words, having missed payment on a specific date the covenant can never be performed exactly in accordance with the lease. This is generally the case with rent, even where the rent payable had been increased retrospectively under a rent review clause (*Parry* v *Robinson-Wyllie* [1987] 2 EGLR 133).

Unfortunately, there is little definitive guidance on this issue either from statute or case law. Each case will be considered on its merits. (The distinction between continuing and non-continuing breaches is also important in considering waiver of forfeiture, as we shall see below.)

Any uncertainty relating to the liability of assignee leaseholders causes problems for landlords, managers and leaseholders (incoming and outgoing) alike. In an ideal world, it should be avoided if all involved in the conveyancing process do their jobs properly. Solicitors and other conveyancers should be making the appropriate enquiries of the property managers — and indeed of their own clients, who need to be forthcoming with information. Property managers meanwhile will serve themselves and their principals best by ensuring that full disclosure is made of any disputes, potential breaches of covenant by the outgoing leaseholder, and particularly any monies which may be due. The latter should include anything that can be foreseen which may need to be added to the account, such as the final cost of ongoing works or any balance likely to be added to the year end service charge account. With that information, the conveyancers can make appropriate retentions from and against the proceeds of sale.

In any event, any contract for the assignment of a lease should contain express indemnities so that if the purchaser faces a claim for a breach committed by his predecessor he can obtain protection.

Relief from forfeiture

Quite apart from the very recent new restrictions on forfeiture, lessees have always been able to take some shelter from a forfeiture action by throwing themselves on the mercy of the Courts of Equity and seeking relief from forfeiture. The history of the conflict and co-operation of common law and equity is long and fascinating, however all that is relevant here is that forfeiture is historically perceived as a harsh common law remedy which has been softened by equitable relief.

As an equitable remedy, relief from forfeiture is theoretically subject to the well-known equitable maxims, such as "delay doth defeat equity" or "he who comes to equity must come with clean hands". However, relief from forfeiture is granted so liberally by the courts that the maxims have long ceased to be applied at all rigorously.

In any event, relief from forfeiture has been enshrined in legislation for some time. In relation to breaches of covenant other than non-payment of rent relief is governed by section 146 of the Law of Property Act 1925; in rent cases the appropriate statutory provisions are in sections 138 and 139 of the County Courts Act 1984.

Relief under section 146

Section 146 provides no forfeiture may be commenced (for breaches other than non-payment of rent — section 146(11)) unless the landlord first serves a notice. The notice is not in prescribed form but it must:

- specify the breach complained of
- require that the breach be remedied (insofar as it is capable of being remedied)
- require the lessee to compensate the landlord for the breach.

Only if the breach remains unremedied and no compensation has been paid within "a reasonable time" (section 146(1)) can the landlord proceed towards forfeiture. What constitutes a reasonable time? This will vary according to the nature of the breach, the capacity of the lessee and so forth. Thus, there is no hard and fast rule. A normal period would be anything from 14 days to a month or even longer in cases, for example, where structural works are required to remedy the breach.

Compensation to the landlord is generally assumed to mean the payment of the landlord's costs and expenses occasioned by the breach of covenant.

Following receipt of the section 146 notice or, more usually, service of the possession claim the lessee is entitled by section 146(2) to apply to the court for relief from forfeiture. Relief is in the court's discretion, but is nearly always granted except in the most extreme cases. Relief has been granted for example even when the breach concerned was an illegal act. However, the court can grant relief conditionally upon terms providing for the breach to be remedied (and any other subsisting breaches) and for steps to be taken to prevent a repeat or continuation.

The court may also award compensation to the landlord, most especially for the payment of his costs and expenses. The landlord's entitlement to costs is specifically provided for in section 146(3), but most will be contractually entitled through the lessee's covenants in the lease (see Chapter 1). Previously such costs were awarded on an indemnity basis which could be out of all proportion to the losses suffered. This practice has been eroded substantially however, especially in the last 10 to 15 years (see below).

Section 146(4) permits any mortgage lenders, subtenants or others deriving title under that of the lessee to make their own applications for relief from forfeiture. Most commonly, such applications have been made by mortgagees whose security will be lost if the lease is forfeited.

Usually they will apply on their borrower's behalf, pay what is outstanding (including costs) and add that to the mortgage debt. Occasionally, in extreme circumstances, they will take a new lease in their own right after the lessee has been removed. In the case of a subtenant, section 146(4) expressly provides that they may not be granted a longer term than under their existing tenancy.

Relief under sections 138 and 139 County Courts Act 1984

These sections apply to forfeiture for non-payment of rent (including for monies reserved as rent by the lease) and thus not to cases where a section 146 notice was served. Section 138 relates to forfeiture by possession proceedings and section 139 comes in where a landlord has effected forfeiture by peaceable re-entry.

By section 138(2) the way of dealing with a case of forfeiture for rent is apparently straightforward:

> If the lessee pays into court or to the lessor not less than 5 clear days before the return day all the rent in arrear and the costs of the action, the action shall cease, and the lessee shall hold the land according to the lease without any new lease.

"Return day" means the date fixed for the hearing. Indeed, this will work very smoothly if both parties can agree the figures without a hearing; however, this is not always so easy especially in relation to costs. Furthermore, section 138(6) clarifies that such a definite conclusion cannot be accommodated if there are outstanding breaches other than non-payment of rent to be considered. In such a case the court may have to consider relief under both section 138 and section 146.

If section 138(2) does not see an end to the case, section 138(3) gives the lessee a second chance by providing that the court may suspend the effect of a possession order for at least four weeks and on terms that it will not take effect at all if the lessee pays the arrears and costs within the period of suspension. Section 138(4) allows further suspensions.

There is one last safety net for the lessee in section 138(9A). This empowers the court to grant relief from forfeiture at any time within six months even from the point at which the landlord takes possession under a warrant. Section 138(9C) extends the right to apply for relief to any mortgagees, subtenants or others deriving title under the lessee.

In the event that a landlord has forfeited a lease for unpaid rent by peaceable re-entry, section 139 of the County Courts Act 1984 gives the court the power to grant relief on application by a lessee, mortgagee or subtenant within six months of re-entry.

Because of this six month window, it is normally prudent for a landlord who has recovered possession because of rent arrears to wait at least six months before entering into any significant dealing in the property, because the lessee can have his lease restored even after a new lease has been granted (*Fuller* v *Judy Properties Ltd* [1992] 1 EGLR 75) and the complications can be horrendous. The same applies with a mortgagee seeking relief (*Bank of Ireland Home Mortgages* v *South Lodge Developments* [1996] 1 EGLR 9).

Costs

Until the early 1990s, it was the practice for landlords to be granted a full indemnity for their costs as part of the terms for relief from forfeiture, even if the breach leading to the use of forfeiture was relatively trivial and notwithstanding the fact that the court has full discretion in costs issues. There were sound reasons for this: landlords were contractually entitled to be paid their costs; costs were also seen as part of the compensation for the breach and thus outside the usual litigation yardstick of "costs of the action"; and the general rule when granting equitable relief is to seek to put the plaintiff back in the position he would have occupied had the breach never occurred.

Indeed, there are still in existence a number of authoritative cases for contractual and/or indemnity costs to be the starting point: see for example *Egerton* v *Jones* [1939] 3 All ER 889, *Factors (Sundries) Ltd* v *Miller* [1952] 2 All ER 630 and *Three Stars Property Holdings* v *Driscoll* [1986] CA.

However, judicial criticism of the practice began to be felt forcefully following Lord Templeman's comments in *Billson* v *Residential Apartments Ltd* [1992] 1 All ER 141, and it has never been quite the same since. At the same time, costs covenants in leases have tended to be construed much more narrowly (see, for example, *Contactreal Ltd* v *Davies* LTL 17/5/01). As a result, it is now much less likely that a landlord will recover his full costs in a forfeiture case, let alone actions for other types of enforcement (such as debt recovery which is subject to such limits on costs awards as those found in small claims) and cases before the LVT (see Chapter 3).

Waiver of forfeiture

Historically seen as a draconian remedy, forfeiture has almost always carried with it a number of obstacles to make it less than straightforward to exercise. One of the oldest is the doctrine of waiver. A landlord may waive his right to forfeit (and thus lose it, at least in respect of the specific current breach) if he does anything which clearly recognises the continuation of the lease. For waiver to occur, the landlord must first have knowledge of the existence of a breach giving rise to the right of forfeiture, but "knowledge" could be deemed by the facts or imputed from others (most obviously, managing agents).

The act of waiver does not have to be deliberate. For example, sending out computer generated invoices will be sufficient; alternatively, serving a statutorily required notice (such as in a consultation procedure under section 20 of the Landlord and Tenant Act 1985) upon a leaseholder against whom forfeiture is contemplated, but in the confident presumption that the leaseholder concerned will obtain relief from forfeiture and thus effectively remain a leaseholder, could be a waiver even if expressed to be "without prejudice" to the forfeiture.

The most obvious act of waiver is to demand or accept rent which has fallen due since the breach. Property managers whose principals are considering forfeiture should ensure they have procedures in place to guard against the possibility of waiver. For example, computer systems may not be able to be prevented from generating invoices for the odd individual property, but it should be possible to "flag" those concerned so that they are not sent out at all or, perhaps, go instead to the property manager's accounts department or the landlord's solicitors (who should know what to do with them).

The period during which there is a risk of waiver runs from the point at which the landlord has knowledge of the existence of a breach until the forfeiture is effected either by peaceable re-entry or service of possession proceedings (or, of course, the breach has been remedied). Here, it is necessary to distinguish between the issue of proceedings which are "legal process" and the service of a notice under section 146 of the Law of Property Act 1925, which is to all intents and purposes a letter before action. The risk of waiver continues after service of a section 146 notice.

There is some solace for landlords however if the nature of the breach is "continuing" (see above). Then, the right to forfeit will immediately re-arise even after a waiver so long as the breach subsists.

None the less, any costs which have been incurred up to the time of a waiver are likely to be lost. It is therefore quite likely that a landlord or manager who falls at this hurdle will face a bill for costs which is higher than the amount of the leaseholder's arrears originally at stake.

Statutory restrictions on forfeiture

Notwithstanding the restraints upon landlords introduced on the courts' own initiative (such as waiver) or the statutory regime for relief from forfeiture (in section 146 and the County Courts Act 1984) Parliament has continued to be troubled by the perceived imbalance towards landlords provided by forfeiture, especially in the light of the extent to which the power has been seen to be abused by "unscrupulous" landlords. Consequently, additional restrictions have been introduced by legislation in recent years. The most effective new measures are contained in section 81 of the Housing Act 1996 and sections 167 to 171 of the Commonhold and Leasehold Reform Act 2002.

The effects of these changes are referred to above and set out more fully in the remaining chapters, but the biggest single impact will be felt over time by the creation of a role in enforcement (albeit indirectly) for the LVT and the consequential change in focus for dispute resolution in residential leasehold from the court to the LVT, the subject of the next chapter.

Summary

It is clear that forfeiture is no longer the weapon it was even 15 years ago. A property manager considering advising his client to embark upon a forfeiture action now in a residential context must bear in mind:

- only the freeholder or immediate landlord can take forfeiture proceedings
- no section 146 notice can be served unless the breach is admitted or there has been prior determination (probably by the LVT)
- with rent arrears, no forfeiture can be commenced for small or short-term arrears
- the leaseholder will almost certainly be granted relief from forfeiture — if not, his mortgagee or sub-tenant probably will
- there is a risk of waiver
- there is likely to be a shortfall on costs recovery.

In the circumstances, it is not surprising that forfeiture is becoming a less popular option all the time.

3 Leasehold Valuation Tribunals: New Jurisdictions and Procedures

Jurisdiction

As its name suggests, the LVT was created principally to engage in the valuation of leasehold properties for enfranchisement and related purposes. The role of the LVT has now greatly exceeded that and it is rapidly replacing the county court as the principal dispute resolution forum for parties to residential leaseholds. The following areas are among those added to the LVT's purview by legislation in the last 10 years:

- reasonableness of service charges and whether they are payable
- similarly with administration charges
- validity of consultation procedures
- appointment of managers
- variation of leases
- determination of disputes regarding right to manage claims
- determinations of breach of covenant.

From the point of view of enforcement of leaseholders' covenants, it is the last of these which is the most directly relevant, although the others have some part to play.

Limitations on jurisdiction

The tribunal's jurisdiction is excluded only in a limited set of circumstances, which can be summarised as follows:

- the question has already been agreed or admitted by the leaseholder (if that is relevant to the circumstances of the application, and see in relation to section 27A referred to below)
- the question has been or is to be referred to arbitration under a post-dispute arbitration agreement (see Chapter 12)
- the question has already been or is to be determined by a court.

Determinations of breach of covenant

Section 168 of the Commonhold and Leasehold Reform Act 2002 (the 2002 Act) provides with effect from 28 February 2005 that no notice under section 146 of the Law of Property Act 1925 may be served unless:

- the leaseholder has admitted the breach of covenant complained of or
- the court or an arbitration under a post-dispute arbitration agreement (see Chapter 12) has finally determined that the breach has occurred or
- the landlord has applied successfully to the LVT for a determination that the breach has occurred.

Section 170 of the 2002 Act brings into line the restriction against forfeiture for non-payment of service charges contained in section 81 of the Housing Act 1996, so that the same stipulations will apply effectively to any conceivable breach of covenant except non-payment of ground rent.

What does this mean in practice?

Assuming that it is very unlikely that leaseholders will be queuing up to admit their breaches of covenant, most landlords who elect to exercise or even threaten forfeiture will first have to apply to the court or the LVT to determine the existence of a breach. (As we shall see later, arbitration clauses in leases no longer provide a viable alternative; the only effective arbitrations will be under agreements by the parties to arbitrate once the dispute is known about.)

Section 168(4) says that a landlord "*may*" make an application to the LVT for a determination, and section 168(2) allows for a determination by "a court in any proceedings" to meet the criteria for forfeiture, so there is no element of compulsion in the Act to apply only to the LVT.

In practice however, unless the landlord is making substantive applications for other relief (an injunction in a nuisance or repairs case for example), it is very likely that the court will take the view that the application would be more appropriately made to the LVT and transfer it. That will lead to delays at best. It may also be that the court will penalise the landlord for issuing in the less appropriate forum by awarding costs against him, or at least preventing him from recovering his costs of the court proceedings.

All in all, unless there are special reasons for going to court, most landlords will be advised to apply straight to the LVT for determinations.

This rule applies only to forfeiture. Any other applications (such as straightforward debt recovery) should continue to be made to the county court.

Interpretation powers

Before the 2002 Act was enacted, the LVT could only determine whether and to what extent a service charge was "reasonable". Occasionally a tribunal would take it upon itself to consider the legal validity of a charge, arguing that only by doing that was it truly feasible to determine if it was reasonable. Generally however the tribunals tended to construe their powers restrictively in this context. That was changed by section 155 of the 2002 Act.

Section 155 inserted a new section 27A into the Landlord and Tenant Act 1985 which gives the LVT power to determine whether a service charge is "payable" and, if so, by whom, to whom, how much and when. Inevitably, in considering whether a charge is payable, the tribunal must satisfy itself that the charge is permitted by the lease and by statute and that all necessary procedures have been followed. Accordingly, to fulfil this function, the tribunal will have to interpret the legal meaning of leases and statutes to a far greater extent than before.

Moreover, section 27A effectively overturned the decision in *R (on the application of Daejan Properties Ltd)* v *London Leasehold Valuation Tribunal* [2001] 3 EGLR 28 (which held that the LVT had no jurisdiction in respect of a charge which had already been paid) by way of section 27A(2) ("subsection (1) applies whether or not any payment has been made") and section 27A(5) ("But the tenant is not to be taken to have agreed or admitted any matter by reason only of having made any payment").

Procedures

The procedures of the LVT have developed somewhat over the years. Much of the current rules and regulations are set out in schedule 12 of the 2002 Act, the Leasehold Valuation Tribunals (Procedure) (England) Regulations 2003 (SI 2003 No 2099) and the Leasehold Valuation Tribunals (Procedure) (Amendment) (England) Regulations 2004 (SI 2004 No 3098). Very helpful guidance notes on procedure are published by the Residential Property Tribunal Service *www.rpts.gov.uk.*

The procedures are simple and straightforward compared to court proceedings; indeed that is the intention. Tribunals should be relatively informal and easy to use to encourage access and to discourage the unnecessary use of lawyers. The simplicity also enables the government to exclude public funding of legal representation.

Although there are some variations because of the increasingly wide range of cases now within the LVT's jurisdiction, there are fundamentally three stages to an LVT case:

- the application
- any directions issued by the tribunal
- the hearing.

The application

Each application should be made on the appropriate form (available from the regional tribunal offices or downloadable from *www.rpts.gov.uk*) and such supporting documentation as is required by the form and indicated by the accompanying guidance notes. In the case of an application for determination of service charge liability under section 27A of the 1985 Act (the most likely option in an enforcement case) the only other document required is a copy of the relevant lease.

The application form requires a brief outline of the issues but not supporting documents or submissions; to that extent the procedure is less "front-loaded" than before and more like the claim form in court proceedings. As with a claim form, the application is to be verified by a "statement of truth" (see Chapters 5 and 7).

A copy of the application is then sent by the LVT to each named respondent. The LVT may also send copies to any other person it considers appropriate (such as other leaseholders at the property who are not named in the application as respondents) with information concerning how the recipient may apply to be joined to the proceedings.

Fees

A fee is payable with the application. This ranges currently from £50 to £350 depending upon the amount of the service charge leading to the dispute. (There is a further fee of £150 payable in the event that the matter is listed for a full hearing.) If multiple applications are made simultaneously the fee payable will be the largest required for any of the applications being made. If the case has been transferred from the court the fee payable will be the same save that the amount of the court's issue fee will be regarded as having been paid as though on account.

Unless fees are waived (very unlikely in an enforcement case) the LVT will proceed only on their payment. Non-payment for a month will be taken as withdrawal of the application.

The Tribunal's directions and case management powers

It is possible to have a determination without a hearing, but only if all parties agree and the tribunal considers it appropriate in the circumstances of the case. The matter can be dealt with on the basis of the documentation and written representations. Obviously this will save the parties considerable time and expense but there are disadvantages, most notably in the lost opportunity of cross-examination. If the applicant considers this approach desirable there is a question on the application form by which it may be requested, although it can be raised later.

Alternatively, if a hearing is necessary, the case might be allocated to the "Fast Track", but only if it is very straightforward and can be dealt with swiftly or if it is urgent (an application for dispensation from consultation for emergency repairs for example). Allocation is up to the LVT but it can be requested on the application form.

Otherwise the case will be allocated to the "standard track". A Tribunal chairman will consider the papers and decide whether a pre-trial review ("PTR") is desirable (if not, either party can request one). A PTR is an informal hearing generally before a single LVT member. Its purpose is not to make any final decisions, but to seek to achieve any or all of the following:

- identify clearly what questions are disputed (or not)
- explore any possibility of settlement on all or any individual points
- consider how the case can be dealt with most expeditiously (including possibly by combining with other cases)
- consider any applications from non-parties to join in
- establish what further steps are necessary to make the case ready to be heard.

Following the PTR or, if none was held, after the initial consideration of the application, the LVT will issue directions to the parties which aim to ensure that all relevant and essential information concerning the facts and the parties' cases is known to the parties and the tribunal. Standard directions include:

- disclosure of information
- written statements of case (these are the parties' submissions)
- provision of copy documents
- production of a paginated and indexed bundle of documents (preferably agreed) for the final hearing
- provision for expert evidence (rarely).

Each point should include a deadline and the directions will also include details of the date and venue for the hearing.

Specifically in relation to disclosure of information, an LVT may give a party written notice to provide the tribunal with such information as it requires by a deadline (not less than 14 days later). Failure to comply is an offence punishable by a fine up to £1,000 at the time of writing (paragraph 4, schedule 12, Commonhold and Leasehold Reform Act 2002).

In exceptional cases a preliminary hearing may be fixed, especially if the LVT's jurisdiction to hear certain matters is brought into question.

By paragraph 7 of schedule 12 to the 2002 Act, the tribunal has the power to dismiss applications (in whole or in part) which are "frivolous or vexatious or otherwise an abuse of process of the tribunal". Dismissal can only be made on notice and the applicant has a right to a hearing. (Further guidance on the way in which the LVT can use this power was given in the Lands Tribunal in *De Campomar and Another* v *Trustees of the Pettiward Estate* [2005] 15 EG 124).

The hearing

The final hearing will usually be before a full panel of three tribunal members who are most likely to be a lawyer (commonly the chairman), a surveyor or valuer and a lay person.

Very often (but not always) the tribunal members will carry out an inspection of the property before the hearing. The parties and their representatives are entitled to be present and to draw the tribunal members' attention to details but not to make any representations. For this reason and for the convenience of the parties, the hearing will be fixed for a venue which is reasonably accessible and close to the property (in London hearings generally take place at the LVT's offices at 10 Alfred Place, WC1).

LVT hearings are relatively informal but none the less are subject to a certain order and rules of conduct which are not entirely dissimilar to court hearings, especially in the "small claims court" for example. The applicant will put his case first and each side will be given opportunities to ask questions and make submissions. The tribunal members may also ask questions of the parties and their witnesses. Where legal authorities or "skeleton arguments" are put forward these should be filed and circulated well before the hearing.

The decision will be discussed by the tribunal members and then issued in writing some time afterwards. Very rarely the decision will be given at the hearing, although occasionally there will be a verbal indication.

Costs

The powers of the LVT when it comes to awarding costs as between the parties remain extremely limited, but they are slightly greater than before the 2002 Act was implemented. A fuller discussion of this subject is found in Chapter 13, but essentially the circumstances in which a tribunal might make an order for costs are as follows:

- by paragraph 9 of schedule 12 to the 2002 Act the tribunal may require one party to reimburse all or part of another's expenditure on the LVT's fees

- by paragraph 10 of schedule 12 the tribunal can decide that one party should pay another's costs, but only up to £500 and only if that party has acted "frivolously, vexatiously, abusively, disruptively or otherwise unreasonably in connection with the proceedings"
- by section 20C of the Landlord and Tenant Act 1985 the tribunal may prevent a landlord or manager from adding the costs of the proceedings to the service charges for the property concerned. (Similar applications may be made in the county court, the Lands Tribunal or arbitrations.)

Appeals (see also Chapter 12)

LVT decisions may be appealed only with permission of the LVT or, failing that, the Lands Tribunal. An application to the LVT for permission to appeal must be lodged within 21 days of the decision being sent to the appellant. It is worth keeping the envelope containing the decision to show the postmark as it may not be the same date as that appearing on the decision itself.

Enforcement of tribunal awards

In the event that a sum of money is awarded by the LVT (which will only apply in limited sets of circumstances), the award may be enforced in the county court under rule 70.5 of the Civil Procedure Rules (CPR). An application for enforcement is made:

- (usually) without notice
- to the court where the debtor lives or trades
- on the straightforward county court form prescribed for the purpose of registering the award (currently N322A)
- accompanied by a copy of the award;

The application can be dealt with by a court official without a hearing. The award can then be enforced in the same way as a county court judgment.

Other orders or directions of the LVT will normally be enforced in accordance with the statute under which they are made; orders relating to right to manage cases for example are enforceable under section 107 of the Commonhold and Leasehold Reform Act 2002.

Where the LVT has determined a question in proceedings transferred to it by the court for that purpose, the court will normally stay or adjourn generally its own proceedings. The applicant can then apply to the court (a letter will usually suffice) to lift the stay or adjournment and list for a short hearing so that the court can effectively incorporate the tribunal's decision within an order.

4 The Role of the County Court

Introduction

Increasingly over the last few years the property manager has had to get to know the Leasehold Valuation Tribunal as more and more of the court's jurisdiction has been pushed in that direction, but the county court still has a major part to play. It is likely that a property manager will spend at least some of his or her time in the county court office or before the district judge.

Because the Human Rights' legislation requires that only exceptional cases will be heard in private now, the distinction between chambers and open court hearings has become less significant. Most civil claims are now open to the public and, thus, are no longer "in camera"; possession cases however are generally in private. For practical purposes most civil hearings are before the district judge and he or she sits the vast majority of the time in his or her hearing room (what used to be called "chambers") rather than in a traditional court room with the judge sitting on high. Inevitably therefore, disinterested members of the public are allowed in only to the extent that there is sufficient space.

Property managers attending court hearings then will most frequently appear before the district judge in his or her own room. No particular rules apply as to dress but due respect should be shown to the Judge, including in terms of appearance. The district judge is addressed as "sir" or "ma'am" (according to gender).

Rules are much stricter in trials in the main court room. Both the judge and the professional advocates generally wear robes and wigs (although this is all a matter for the judge; some eschew wigs for example). Some possession hearings are still heard in the main court room, especially if there are several witnesses.

The judiciary

District judges have jurisdiction on all possession claims and in money claims up to £15,000. Higher value cases go to the circuit judge or High Court judge (depending upon value and sometimes location). Appeals from the district judge in the county court are heard by the circuit judge. Appeals from the circuit judge either go straight to the Court of Appeal or to a High Court judge depending upon the type of case. Recorders (part-time circuit judges) sometimes sit on cases which are appropriate for the circuit judge. Circuit judges and recorders are addressed as "Your Honour".

Some quasi-judicial functions are now undertaken by the court clerks, particularly in the enforcement context.

Lawyers

The first point to make is that there are now very few circumstances in which it is compulsory to be represented by a lawyer. It is almost impossible to think of such a scenario in an enforcement case. An experienced property manager should be able to deal with most cases, especially if there is no substantive defence or no complex point of law arises. None the less, not everyone has the gift of advocacy (indeed this is as true of lawyers as anyone else), and there may be situations when it is in the client's interest to instruct a legal professional. This can be just in the context of advocacy or in preparing the client's case in pleadings or submissions, or in respect of the whole proceedings. Normally, if a lawyer is instructed, they will file and serve the documentation on the client's behalf and will appear on the court record for that party. If the lawyer is disinstructed or has only limited instructions they must inform the court of the position.

Barristers (or "counsel") are mostly advocates who specialise in advocacy and generally in particular areas of law. Although some of the old rigidity has been relaxed it is still typical for barristers to be instructed through solicitors. It is common now however for barristers to appear without their instructing solicitors needing to be present, especially in relatively quick or straightforward cases. Barristers can be instructed by other professionals (chartered surveyors for example), but the barristers may want reassurance first that they are covered by the professional indemnity insurance of those instructing them. Negotiations will be needed with the barristers' clerks.

The majority of barristers are self-employed individuals who gather together in groups of chambers. Their diaries, accounts and so forth are handled by their clerks. Solicitors who frequently go to a particular set of chambers are in a better position to negotiate a good deal with counsel's clerks so it can be a false economy to leave the solicitors out of the equation.

Barristers are entitled to appear before any court in the land. Senior barristers (Queen's Counsel or "silks") tend to handle the most complicated and highest value cases, particularly on appeal. A QC will generally be supported by a junior barrister.

Solicitors tend to form partnerships and offer much more comprehensive legal services than barristers. A solicitors' firm will generally consist of a range of fee earning staff charging a variety of different rates. Although some solicitors still practise individually, the trend has been for larger and larger firms with specialist departments (relatively few practise specifically in the residential leasehold field however).

Partners in solicitors' firms are the most experienced and the most expensive. Most day to day file handling will be carried out by employed staff who may be assistant solicitors, legal executives, trainees of either ilk, or unqualified fee earners (generally known as paralegals, although many paralegals have legal qualifications or are in the process of obtaining them).

Only fully-fledged Fellows of the Institute of Legal Executives are strictly authorised by their Institute to use the description "Legal Executive". Legal executives must be supervised by solicitors in the conduct of litigation, although the degree of supervision relaxes considerably as experience and seniority are gained. It is not now uncommon to find legal executives leading substantial departments, often supervising assistant solicitors in their turn.

Legal executives are not entitled to the same rights to appear in court (rights of audience) as solicitors, but they are entitled to conduct cases before the district judge.

Debt recovery

From the enforcement point of view, the greatest relevance of the county court is in terms of debt recovery. Since the introduction of the

Civil Procedure Rules 1998 (CPR) the county court has had a system of allocating cases to one of three tracks. Generally, allocation will be determined by the amount of the claim, as follows:

- small claims track — up to £5,000
- fast track — £5,000 to £15,000
- multi-track — over £15,000.

Allocation only crops up if a case is defended however, and it can be influenced by a number of factors, including complexity, volume of evidence and public interest.

District judges have jurisdiction to try cases in the small claims and fast tracks; circuit judges tend to try multi-track cases. The scales of costs awarded by the court also vary according to the track.

Conducting a debt claim in the county court is examined in detail in Chapter 7.

Possession orders

Virtually all possession claims are heard in the county court by district Judges. The procedures differ substantially from debt claims as does the question of allocation to track.

Possession proceedings are looked at in the context of forfeiture in Chapter 5.

Relief from forfeiture

The county court also has an inherent equitable jurisdiction to grant relief from forfeiture either on application during possession proceedings or on a free-standing claim.

Other remedies

The county court possesses a wide range of other weapons in its armoury for the disgruntled litigant to avail himself. The following are examples of orders which a property manager might seek when enforcing covenants:

- declarations (for example that a service charge is validly due as a preliminary step to forfeiture)
- injunctions (restraining alterations or nuisance behaviour for instance)
- decrees of specific performance (of covenants)
- leave to bring a forfeiture claim for breach of repairing covenant under the Leasehold Property (Repairs) Act 1938 (see Chapter 8).

Claims for remedies other than financial awards are dealt with differently by the courts. Whereas money claims are issued under part 7 of the CPR, most of the above will be made under part 8 especially if they are not combined with financial claims. District judges have jurisdiction for nearly all part 8 claims no matter the notional value of the claim.

The procedures for such claims are considered in Chapters 8 and 9.

Costs

It is undoubtedly more difficult to recover a party's full legal costs in the county court than it used to be. There are a number of reasons for this.

- The court has always had complete discretion on costs issues, but until relatively recently judges were reluctant to interfere with contractual provisions allowing landlords to recover full indemnity costs. That has changed, partly because of case law (see Chapter 2), but partly as a result of the Civil Procedure Rules.
- CPR part 1 sets out "The Overriding Objective" which requires (among many other things) that cases are dealt with proportionately. Where cases are perceived as straightforward or involve relatively small sums of money proportionality tends to be applied somewhat zealously to the levels of costs involved.
- Generally in all small claims track cases, only pre-set fixed costs will be allowed; similarly in proceedings where judgment is obtained in default. The fixed costs regime is extremely restrictive.
- The increase in the small claims limit to £5,000 took huge numbers of cases out of reach of costs recovery. Very few management companies can afford to let arrears build up to over £5,000 before taking proceedings.
- The fast track (for cases between £5,000 and £15,000) also has a structure of fixed costs, although rather more generous than in the small claims track.

- The court has the power (which judges are encouraged to use) to conduct "summary assessments" of the costs of the party in whose favour costs are awarded. These generally take place at the end of a hearing rather than as a separate hearing and of necessity they tend to be rough and ready. Summary assessments do have the distinct advantage however that they have drastically cut the need for long drawn-out (and expensive) taxation processes. Taxations (or detailed assessments as they are now known) are still needed in more complex actions, after lengthy trials or in a few other circumstances, which are outside the scope of this book.
- Generally, the levels of costs awarded by the courts have failed to keep pace with the increases in solicitors' charging rates or barristers' fees.

Ground Rents

5

Notices and demands

As we have seen in Chapter 2, ground rent now has to be demanded in accordance with the prescribed form before payment can be enforced (section 166 of the Commonhold and Leasehold Reform Act 2002; the form is found under The Landlord and Tenant (Notice of Rent) (England) Regulations 2004 SI 2004 No. 3096). This provision came into force on 28 February 2005. At the time of writing it was too early to see what consequences this led to in terms of additional administration cost, to what extent the regulations have been followed, and the practical effects of any failure to comply.

What can be easily anticipated however is that in the majority of cases non-compliance discovered subsequently will have expensive results in dismissed and abandoned proceedings for forfeiture or debt recovery. It is therefore essential that notices be issued correctly.

Due dates

The due date for payment of rent set out in the lease diminishes in relevance unless the notice is issued in time for it to be effective, as the due date will now be no earlier than 30 and no later than 60 days after the notice has been given and, in any event, it cannot be before the due date in the lease.

For example, if the rent is due on 25 March under the lease, notice must be given by 23 February at the latest for it actually to be due on that date. Notice could be given as early as 22 January, but no earlier.

Because section 166 is so specific in its requirements, it seems unlikely that any compromise formula could be reached by, for example when rent is due quarterly, issuing one annual ground rent notice with quarterly reminders. Obviously the leaseholders could agree to such an arrangement but if any defaulted in payment proceedings against them would almost certainly fail. Even one defaulter could thus render such a scheme counter-productive from a cost-effective point of view.

Service

A certain amount of litigation can be expected to clarify the precise intention of section 166 when it comes to service of ground rent notices. Section 166(1) absolves a leaseholder of liability to pay rent "unless the landlord has given him a notice", but nowhere is "given" defined. The interpretation of "given" has already caused some disquiet in the LVT in the context of consultation notices under section 20 of the Landlord and Tenant Act 1985. For example, is a notice "given" until it has been "received"?

Section 166(5) lays down that the notice must be in the prescribed form and it *may* be sent by post (emphasis supplied). If sent by post, section 166(6) requires that it be sent to the leaseholder at the demised property *unless* he has given the landlord written notice of an alternative address in England or Wales for this purpose (in which case the notice must go to that address).

"Sent by post" has the benefit of a statutory definition in section 7 of the Interpretation Act 1978 which tells us that service of an item, properly addressed, pre-paid and posted, is deemed to be effected at the time at which it would be delivered in the ordinary course of post (two working days after posting for first class post) unless it is proved not to have been delivered.

The way the service provisions are framed poses some practical problems for property managers, especially for sublet flats. What if:

- the leaseholder gives no address but the property manager doubts that sending it to the flat will reach him?
- an address is given over the telephone, but is not confirmed in writing?
- the leaseholder supplies a contact address but not expressly for the purpose of ground rent notices?
- the address given is not in England or Wales?

- the address is that of the leaseholder's letting agents or solicitors but, when notices are sent there, they return them declining to accept service?
- the property manager suspects the address is out of date?
- post sent to the address given is returned undelivered?

In any of these situations, it seems as though sending the notice to the demised flat is likely to be a safer approach even if the property manager knows that the leaseholder is not there and strongly doubts that he will receive post sent there. Alternatively, it is permissible to hand deliver the notice through the letter box of the demised property; section 166 (6) only applies if the notice is to be sent by post. Thus, it would appear that personal service at the property will be effective no matter that the leaseholder may have supplied a perfectly sustainable alternative address.

One final point on service of ground rent notices: it was the practice of some landlords to refrain from sending out rent demands to leaseholders who were in breach of covenant (possibly entirely unrelated to rent) in order to avoid any waiver of the right to forfeit (see Chapter 2). This could be done in the comfortable knowledge that rent would continue to accrue whether demanded or not. Obviously that comfort is no longer available, but the fact that sending out a notice is a statutory requirement does not protect landlords from risking waiver. A ground rent notice will waive the right to forfeit, so landlords in this position will have to choose whichever approach is the lesser of two evils in their particular circumstances.

Arrears recovery

In order for rent to be recoverable then it must be due under the lease and it must have been the subject of a notice in the prescribed form, validly served. There is still one more obstacle however, and that is imposed by sections 47 and 48 of the Landlord and Tenant Act 1987. In summary, these sections provide that any demands (including for rent, but also service charge or administration charge) must contain the landlord's name and an address in England or Wales where the leaseholder may serve notices or proceedings upon the landlord. If this information is not supplied, no monies will be due from the leaseholder. Of course it is likely that the ground rent notice will contain the necessary details, but it is not inevitable.

Subject to the necessary formalities having been completed satisfactorily, the property manager needs to consider how to collect in arrears. From the point of view of formal proceedings, the options are debt recovery through the county court or forfeiture (considered below). Distress for rent is hardly ever seen as a realistic or even a savoury alternative nowadays for residential premises. There are other methods of debt collection through factors and other agents, but they are outside the scope of this book. None the less, a prudent property manager will have considered whether such measures might work more effectively in particular sets of circumstances.

Debt recovery through the county court

Chapter 7 will deal with county court proceedings generically, but the issues which are likely to be peculiar to a case for arrears of ground rent include:

- it is extremely unlikely that ground rent arrears on their own will exceed £5,000, so any defended case will be allocated as a small claim
- having said that, it is rare for there to be a meritorious defence to a claim for rent. Examples of defences which might be run include an administrative point on service of notices or a claim for a set-off for, say, the leaseholder's expenses on an urgent repair (which would more properly be classified as a counterclaim)
- a default judgment may be sufficient to elicit payment from the leaseholder's mortgage lender (see below) without any further enforcement action.

Forfeiture for non-payment of rent

In Chapter 2 we have examined the differences between forfeiture for rent and for other breaches of covenant; we also looked at the different ways in which relief from forfeiture can be obtained by the leaseholder. To summarise the basic differences:

- there is no need to apply to the Leasehold Valuation Tribunal to establish that a breach has occurred

- there is no requirement to serve a notice under section 146 of the Law of Property Act 1925 (a "section 146 notice") — indeed none should be served
- relief from forfeiture is sought under sections 138 and 139 of the County Courts Act 1984
- it is also worth remembering that common contractual provisions for costs refer specifically to section 146 so it is unlikely that the landlord can argue for his costs of the proceedings on that basis (as section 146 is effectively irrelevant to issues concerning rent).

From a strategic point of view there is one other significant difference. Because the existence of the breach does not have to be proved and the amount of the rent is fixed by the lease, there is little room for dispute — the rent has either been paid or it has not. Moreover, because so little formality is required before the landlord can proceed to forfeit, the risk of forfeiture is so much more immediate. In the light of the combination of those factors, most mortgage lenders will still clear their borrowers' arrears on an approach from the landlord or his agents. (What is now less certain is whether they will also pay the landlord's costs without question.)

The first step for the property manager therefore will be to check whether the lease is mortgaged and, if it is, apply to the lender to discharge any arrears. If that does not produce payment or if there is no mortgage, it is necessary to consider the formal steps towards forfeiture. Essentially these steps are:

- letter before action (this is effectively required by the courts' Practice Direction on Protocols — failure to serve such a letter will likely lead to penalties, especially in costs);
- issue proceedings for possession out of the county court with geographical jurisdiction for the premises
- attend the hearing to prove the claim and the entitlement to possession and seek an order for possession with payment of the arrears, interest and costs
- request the issue of a warrant of possession and attend with the bailiff
- apply to the Land Registry to close the leasehold title and proceed to grant a new lease of the premises or otherwise deal with it as desired.

Of course it is extremely rare that the leaseholder or some other interested party does not seek and obtain relief from forfeiture, especially in a rent case.

The scenario below is intended to illustrate the proceedings more effectively.

Scenario 1: Reston Laurels plc v A Wohl

(For more details of the location and characters please refer to Appendix 1: Case Study.)

Mr Wohl is the leasehold owner of Flat 5, Tumbledown Mansions (along with two others in the block). Reston Laurels Plc is the freeholder, but the block is managed by Tumbledown Management Ltd through their agents Reddy & Willing. Reston Laurels have an agreement with Reddy & Willing to collect ground rents, which is fortunate because the lease contains a covenant by the leaseholder to notify the management company of assignments (including new leaseholders' addresses) but not the freeholder.

As Mr Wohl is believed to reside abroad, ground rent notices have been sent by Reddy & Willing, in the proper form, to his letting agents "Play A Let", being the correspondence address supplied by Mr Wohl's solicitors on his purchase. The agents are in the same town. Mr Wohl's subtenants are Mr and Mrs T Dance (an elderly couple).

The ground rent for the flat is £100 pa, payable half-yearly in advance on 24 June and 25 December. The freeholder's policy is to take forfeiture proceedings as soon as arrears meet the criteria or one of them under section 167(1) of the Commonhold and Leasehold Reform Act 2002. Section 167(1) states:

> A landlord under a long lease of a dwelling may not exercise a right of re-entry or forfeiture for failure by a tenant to pay an amount consisting of rent, service charges or administration charges (or a combination of them) ("the unpaid amount") unless the unpaid amount —
> (a) exceeds the prescribed sum, or
> (b) consists of or includes an amount which has been payable for more than a prescribed period.

By virtue of the Rights of Re-entry and Forfeiture (Prescribed Sum and Period) Regulations 2004, the prescribed sum is £350 and the prescribed period is three years. The freeholder has received no rent for this flat for four years and thus the arrears stand at £400. Reston

Laurels plc therefore instruct their solicitors; for enforcement purposes they use the same solicitors as the management company, Messrs Swift & Sharp.

The letter before action

Swift & Sharp write a letter before action to Mr Wohl at the letting agents' address and, as a "belt and braces" measure, they send a copy to him as at Flat 5 (which Mr Dance subsequently hands in to the agents unopened). They also conduct a search against the registered title, but this supplies no new information: in particular, it confirms the absence of any mortgage.

The letter before action should comply generally with the requirements of the Practice Direction on Protocols under the CPR with respect to letters of claim. Essentially, it should give a clear and concise summary of the facts and what is sought. Reference should be made to the availability of independent legal advice. The letter should give a reasonable time for a reply and inform the recipient that proceedings will be commenced without further notice if no satisfactory reply is received by a given date. If any information is requested, a reasonable time should be allowed for any necessary investigation and reply.

In this case, the facts to be stated are the names of the parties and property and the basic details of the lease (especially the covenant to pay the rent and the forfeiture clause) and a statement setting out the arrears of rent. The letter should indicate that the landlord will be issuing proceedings for possession of the premises and that the proceedings will also include claims for the arrears, interest and costs. Fourteen days would seem a reasonable period for a reply, although that could be extended to allow comfortable time to reach Mr Wohl, particularly if he is asked to confirm an appropriate address for service (see below).

Issuing the proceedings

When the time given in the letter before action has expired (and presuming that the arrears remain unpaid) Swift & Sharp will advise Reston Laurels that they may proceed to forfeiture by issuing a possession claim in the county court. They know that peaceable re-entry is not possible in this case because the property is still in residential occupation (by Mr and Mrs Dance). Reston Laurels instructs Swift & Sharp to go ahead.

Before even drafting the claim, the solicitors need to be sure that they have an appropriate address for service of the proceedings upon Mr Wohl. (For the purpose of this scenario, we shall assume that it has proved impossible to ascertain Mr Wohl's own correspondence address.) The fact that the landlord and the agents have complied with the 2002 Act as regards service of the ground rent notice does not obviate the necessity to serve court process upon an address which will be acceptable to the court. For straightforward service of the claim by the court, the claim form must show the defendant's address in England or Wales or that of his solicitors. The solicitors' address is only valid if they are authorised to accept service for the purpose of the proceedings. If the claimant cannot supply such an address, application must be made to the court for service by an alternative method. An application must be supported by evidence showing what efforts have been made to establish an address and why some other method is needed. (The court's rules relating to service are set out in rule 6 of the CPR.)

One possible approach for Swift & Sharp in this case is to apply to the court for permission under CPR rule 6.14 to serve the claim upon Mr Wohl's letting agents. This rule allows service on an agent of an overseas principal, but only if the claim concerns a contract made in England or Wales and the agent is still in a business relationship with the defendant. The supporting evidence could include a statement from Mr and Mrs Dance (the subtenants) that they continue to pay their rent to Play A Let — which should establish the second point — but some creative drafting may be needed to argue that the contract was made within the court's jurisdiction. The difficulty here is in determining which is the relevant contract; that is, is it the lease or the assignment to Mr Wohl (to which Reston Laurels was not a party)?

The more usual alternative in the case of a sublet flat is to ask the court's permission to effect service both at the demised flat and by sending a copy to the letting agents. Even here the application will not be routine because the court will still need to be persuaded that reasonable efforts have been made to find a better address for Mr Wohl and that there is a good prospect that the documents will come to his attention if served in this way. When making an order the court will usually extend the time for the defendant to respond to the claim, taking into account that the claim will reach him in a roundabout fashion.

The claim form

There is a prescribed form for claiming possession (N5) and for the particulars of claim (N119). These are available from the county court or they can be purchased from legal stationers. They can also be downloaded from the court service website *www.hmcourts-service.gov.uk*. (A sample claim form and particulars of claim for this case and notes regarding their completion can be found at Appendix 2.)

Care must be taken in completing these forms. From a practical point of view they need some modifying to fit the circumstances of a long leasehold case. Moreover, the person who signs the forms is also committing himself to what is called a "Statement of Truth", verifying the contents of the remainder of the form. It is essential that the right person makes the statement: the solicitor conducting the case is frequently used, but if it is someone from the landlord's ranks it must be a person who can verify the facts from his own knowledge and who truly represents the party (in other words, the managing agent is not the appropriate person). A director or other officer of the landlord company is best suited for this role.

A person who makes a false statement may be liable to punishment for contempt of court (which means imprisonment potentially) so it is worth ensuring that the facts as stated in the documents are accurate.

The claim must disclose full details of the lease and the easiest way to do that is to annexe a copy. It is also usual to annexe an up to date office copy of the entries for the property at HM Land Registry as these will show how and when the defendant acquired the leasehold title and whether there are any registered charges (the holders of which will be entitled to apply for relief from forfeiture (see Chapter 2).

The defendant's response to the claim

It is not compulsory for the defendant to respond to the claim. He will be criticised if he seeks to defend the proceedings late, and will probably be penalised in costs, but he will be heard. In a claim for rent a defence is much less likely, but the defendant would be expected to apply for relief from forfeiture at some stage.

A possession order cannot be entered in default so it will be necessary to have a hearing at which the claim will have to be proved. In our scenario, if Mr Wohl files no response to the claim, the claimant's problem will be to establish on the balance of probabilities that he received notification of the proceedings. In readiness for making out such an argument Swift & Sharp will no doubt be in correspondence with the letting agents asking them to confirm that Mr Wohl has been informed; possibly they will even threaten the agents with a witness summons to answer similar questions at the hearing.

The hearing

As can be seen from the claim form (Appendix 2) the court endorses the form with a hearing date. Although practice varies somewhat from court to court, generally this will be a date when the court is due to hear a lengthy list of possession claims (often regarding assured shorthold tenancies) when each case will have a very short time allowance. In an undefended case this should be the final hearing, so the claimant and his representatives must be ready to prove their claim. They will need the following in order to succeed.

- A duly authorised representative of the landlord, Reston Laurels plc, to give evidence of title and the arrears. The witness could be from Reddy & Willing, the managing agents, but ideally he should be able to produce written authority from the freeholder. (The witness may not be called but he should be there in case the judge requires him to give evidence in person.)
- The original counterpart lease (which should be stamped) should be ready to be produced if required by the judge.
- Swift & Sharp should have obtained up-to-date office copy entries from HM Land Registry to confirm that there have been no dealings in the leasehold title since the claim was issued. Indeed, some judges insist on seeing office copy entries of the freehold title (if the claimant is not the original freeholder) to satisfy themselves as to the claimant's title.
- An up to date statement of arrears should be produced.
- Swift & Sharp should have calculated interest up to the date of the hearing (see Appendix 2).
- A schedule of the claimant's costs up to and including the costs of the hearing should have been prepared at least two days beforehand and an attempt made to serve it upon Mr Wohl. Clearly service will be difficult in this case and the court will not expect Swift & Sharp to have acted disproportionately, but at least copies (addressed to Mr Wohl) should have been sent to the flat and the letting agents. The judge will have been awarding fixed costs on the assured shorthold and secure tenancy cases in the list,

so Swift & Sharp may have to be prepared to push hard for a higher scale, with authorities ready to be cited (see Chapter 2 on the subject of costs in forfeiture actions).

- Finally, the claimant's lawyer will have to be ready to explain how the proceedings have been served and provide evidence of service if it is not already on the court file (and, sometimes, if it is).

Assuming that the judge is satisfied with the evidence and submissions, all that remains is to obtain the possession order together with a money judgment comprising the arrears of rent, *mesne* profits, interest and costs.

Incidentally, the above description of the hearing is on the assumption that Reston Laurels will be represented by its solicitors. It may be that the solicitors instruct counsel (indeed a barrister is frequently a more economical option). Alternatively, there is no legal reason why the managing agent could not represent the claimant so long as the necessary formalities are observed (especially the prior service and filing of a notice that the solicitors are withdrawing from the case — "coming off the record" — and it may be that the judge's permission will be required depending upon local practice).

Recovering possession

The possession order will normally require the defendant to deliver up possession within 28 days (this period can be extended or, in very extreme circumstances, shortened but 28 days is the norm) and the landlord can take no steps until that period has expired. Indeed if, within that period, the defendant discharges the arrears and costs the landlord's right to possession is excluded (see Chapter 2 regarding relief from forfeiture for non-payment of rent). All this is spelt out in the form of possession order issued by the court.

If nothing substantive has been heard after 28 days, Swift & Sharp will request the court to issue a warrant for possession (together with payment of the appropriate fee). This is a paper transaction and does not need to go before the judge; none the less it is important that the request is completed accurately. It is also prudent to inform the court at this stage that the property is occupied by persons other than the defendant (and, in this case, that they are an elderly couple). There is a space at the foot of the prescribed form (N325) to give useful information of this sort.

In due course, the county court bailiff will issue an appointment for executing the warrant. A duly authorised representative of the landlord will need to be present to receive possession from the bailiff and secure the premises, so Swift & Sharp will need to arrange this and ensure that a locksmith is booked.

It is extremely rare for a forfeiture action to proceed to the point where the landlord recovers possession. When it does happen, it is frequently discovered that the leaseholders have already left the premises for whatever reason, and the execution of the warrant is thus extremely straightforward. Occasionally however persons are found in occupation and all the parties concerned need to be prepared for this. If opposition is expected a police presence will have been requested by the bailiff. Sometimes though it is the unexpected which occurs — for example the premises may have been sublet without the landlord's knowledge.

The landlord should have a strategy worked out in anticipation of finding someone in occupation. In our case, it can be anticipated that Mr and Mrs Dance will be present (see below).

After possession

In the event that vacant possession is obtained, the landlord is free to deal with the property as he thinks fit. For example, it could be let on an assured shorthold tenancy or marketed on the basis of a new long lease. Subject to any applications for relief from forfeiture (see Chapter 2) the leaseholder no longer has any interest in the property; nor does he have any claim on the proceeds of sale or any rental income from the flat. The landlord would be prudent to ensure that the leasehold title is closed before implementing any long-term plans. This is achieved by a relatively simple application to HM Land Registry backed by a statutory declaration.

Although the leaseholder's title in the property is extinguished, the landlord still has a claim against him for the arrears of rent prior to the completion of the forfeiture and any interest and costs which have been awarded. In practice, these claims are hardly ever pursued as the value of the flat normally far outweighs the amounts due.

The position of the subtenants

As discussed in Chapter 2, subtenants are among the class of persons who are entitled to seek relief from forfeiture, but they cannot achieve a longer term than they already have. If they wanted to stay living in the flat Mr and Mrs Dance had two options:

- they could apply to the court for relief from forfeiture, or
- they could do a deal direct with Reston Laurels plc to take a new tenancy from them.

The second option is probably the most economic approach, especially if they can negotiate a longer term or other improved terms in consideration for not complicating the forfeiture proceedings. It is not unheard of for such a deal to be struck even as the bailiff attends to execute the warrant, with the new key being handed to the subtenants immediately on the bailiff's departure.

Summary of the new restrictions

In summary then it can be seen that not a great deal has changed in principle regarding forfeiture for unpaid ground rent when it comes to the procedures — they are still relatively simple. What has changed fundamentally is the frequency with which the opportunity will arise. Forfeiture proceedings cannot be commenced unless and until:

- ground rent has been demanded by a valid notice on the prescribed form and
- over £350 is outstanding, or
- some amount has been outstanding for over three years.

Nevertheless the ultimate sanction remains for the worst defaulters or for leaseholders who simply disappear or abandon their flats.

Restrictions in relation to the landlord's nominated insurer

Coming into force at the same time as the restraints on forfeiture for ground rent were the new rules affecting leaseholders required by their leases to insure through the landlord's nominee. Parliament equated those landlords who forfeited leases for trivial rent arrears with those who were perceived to take advantage of such insurance provisions. The new insurance rules relate expressly to leasehold houses.

Section 164 of the 2002 Act provides that the covenant to insure through the landlord's nominated or approved insurer is of no effect if:

- the house is insured through an authorised insurer (one who does not contravene the prohibition contained in section 19 of the Financial Services and Markets Act 2000)
- the policy covers the interests of both the landlord and the leaseholder
- the policy covers all the risks required to be covered by the lease
- the policy covers at least the sum assured as required by the lease
- the leaseholder serves notice of the cover as required by section 164(3).

Under section 164(3) the leaseholder is to give a "notice of cover" to the landlord within 14 days from the existing renewal date or, if it has not been renewed, the date the cover took effect. If the freehold changed hands, an incoming landlord may request a notice of cover within a month of his acquisition, and the leaseholder must supply it within 14 days of the request.

The notice of cover should be in the form prescribed by the Leasehold Houses (Notice of Insurance Cover) Regulations 2004 and supplemented by the Leasehold Houses (Notice of Insurance Cover) (Amendment) Regulations 2005, and it must contain:

- the name of the insurer and the policy number
- the covered risks
- the amount and period covered
- the policy excess
- statements that the leaseholder is satisfied that his interests are covered and that he has no reason to believe that the landlord's interests are not covered.

The notice may be sent by post and, if it is, it must go the landlord's address last furnished to the leaseholder under section 48 of the 1987 Act or, failing that, to the last address furnished under section 47 of the 1987 Act. However if the landlord has supplied a specific address in England or Wales for service of notices of cover that is the address which must be used (section 164(9)).

6 Service Charges: The Options

We have seen in the previous chapters that the Commonhold and Leasehold Reform Act 2002 has had a variety of effects upon enforcement of covenants, including covenants to pay service charges. The purpose of this chapter is to focus on the various ways in which service charges recovery may be maximised in the light of the legislation.

Given that the 2002 Act changed the definition of "service charge", it may be worth repeating that definition as it now stands (section 18, Landlord and Tenant Act 1985 as amended):

> (1) In the following provisions of this Act "service charge" means an amount payable by a tenant of a dwelling as part of or in addition to the rent
> (a) which is payable, directly or indirectly, for services, repairs, maintenance, improvements or insurance or the landlord's costs of management, and
> (b) the whole or part of which varies or may vary according to the relevant costs.
>
> (2) The relevant costs are the costs or estimated costs incurred or to be incurred by or on behalf of the landlord, or a superior landlord, in connection with the matters for which the service charge is payable.
>
> (3) For this purpose —
> (a) "costs" includes overheads, and
> (b) costs are relevant costs in relation to a service charge whether they are incurred, or to be incurred, in the period for which the service charge is payable or in an earlier or later period.

The salient points which can easily be missed are that insurance contributions are service charges (however they are defined in any particular lease) and the cost of improvements is now included as well.

Administration charges are not service charges as such, although they are now treated in a very similar way.

Service charges reserved as rent

To begin with, a distinction needs to be drawn between service charges which are reserved by a lease as additional rent and those which are not. The same points will apply where insurance is treated separately by the lease, often categorised as "insurance rent". In practice, these distinctions are far less important than they were.

The only significant issue is that notices under section 146 of the Law of Property Act 1925 ("section 146 notices" — see Chapter 2) should not be served for rent or monies reserved as rent (which includes service charges and insurance contributions described in that way in a lease). As a matter of good practice, some formal letter before action should be sent to leaseholders if forfeiture is proposed even if it is not a section 146 notice.

Save for the service of a section 146 notice, the procedures for forfeiture for non-payment of service charges are effectively the same, whether they are reserved as rent or not.

Forfeiture via the LVT

Of course, the LVT cannot forfeit a lease or order possession of a property; that power remains with the court and is likely to continue to do so. However, the LVT has a crucial role in forfeiture as it is now unlawful to commence forfeiture proceedings for service charges unless:

- the breach of covenant giving rise to forfeiture is admitted by the leaseholder, or
- its existence is determined by the court or the LVT or by arbitration.

(Section 170 Commonhold and Leasehold Reform Act 2002.)

Arbitrations will only be effective by way of a post-dispute arbitration agreement (which is unlikely to be a popular choice in most cases) and it is highly likely that most county court judges will transfer service charge cases to the LVT. Consequently it can be seen that most service charge arrears cases will almost inevitably involve the LVT at some point if forfeiture is the landlord's preferred option: see Chapter 3 for further discussion on these points.

There is one obvious advantage to going through the LVT for a determination, particularly in larger estates. The LVT can deal with the same case regarding a number of different leaseholders in a single application (so long as the leaseholders all contribute to the same service charge fund) whereas the court cannot. The decision in such proceedings may then be used as the basis for as many individual forfeiture actions in the county court as may be necessary.

The fundamental procedures for an LVT case are described in Chapter 3, so the following scenario is intended simply as an illustration of the way in which a case with multiple respondent leaseholders might develop.

Scenario 2: LVT case regarding Tumbledown Mansions

(For more details of the location and characters please refer to Appendix 1: Case Study.)

Tumbledown Management Ltd wishes to collect service charge arrears from the leaseholders of Flats 3, 7, 10, 12 and 17. The directors discuss with Reddy & Willing (the managing agents) how to go about collecting their arrears and, in particular, whether forfeiture or debt recovery is likely to work best. After some deliberation it is decided to take out one LVT application regarding all five and keep the forfeiture option open for all of them.

(When it comes to forfeiture, only the freeholder Reston Laurels plc can take the proceedings as they are the party entitled to seek possession. As the LVT is only determining what is payable Tumbledown Management Ltd can be the applicant. From a "belt and braces" point of view however it may be prudent for Reston Laurels plc to be a joint applicant so that there is no question subsequently about their entitlement to claim forfeiture based upon a decision in proceedings to which they were not party. By the same token, the freeholder should not be the only party to the LVT application as Reston Laurels are not contractually entitled to receive service charge payments.)

An application form and very helpful guidance notes can be downloaded from *www.rpts.gov.uk*, the website of the Residential Property Tribunal Service (which includes the LVT) or obtained from RPTS panel offices. The form is completed (see Appendix 3) and submitted together with the fee and a copy of the lease. Although this can be clarified and confirmed later, it is simplest to supply a copy of one of the leases concerned and verify in the covering letter that the service charge provisions in all the leases are identical (presuming that they are of course). Apart from the papers mentioned already, no other documents should be supplied at this stage. The parties appoint Reddy & Willing as their representative to begin with, but agree that they will jointly appoint Swift & Sharp solicitors to act depending on how the case develops and particularly what responses are received.

The LVT proceedings

The respondents' actions on receipt of the application vary considerably.

- Mr Hickok and Miss Oakley of flat 3 write in to the LVT (the clerk copies their letter to Reddy & Willing) saying that the reason they have not paid is that they have consistently suggested that Wild Bill Builders (Mr Hickok's uncle) could have carried out repairs and maintenance at a fraction of the cost actually incurred.
- Miss Keen of flat 7 is concerned by the prospect of her employers learning that she is a respondent to proceedings so she sends Reddy & Willing a five page letter (which she copies to the LVT) accompanied by a cheque for her arrears. In her letter however she asserts that she is paying under protest and reserves the right to dispute the legality of some of the charges.
- Mr Straw of flat 10 writes to Reddy & Willing (copied to the LVT) saying that he will pay what he owes as and when he sells his flat, but that he is not minded to pay earlier because of the management company's lack of action regarding his complaints — especially about noise from the flat above.
- Neither of the leaseholders for flats 12 or 17 make any response.

Given the variety of responses and the number of parties, the LVT decides to hold a pre trial review (PTR) (see Chapter 3).

Ms Reddy of Reddy & Willing attends the PTR which is before a tribunal chairman sitting alone. None of the respondents attends.

The Tribunal Chairman, a solicitor (Mr Evan Hand), is concerned that a number of points raised by the respondents so far indicate the possibility of convoluted legal argument at the final hearing. The letter from Mr Hickok and Miss Oakley suggests there may be a dispute over consultation under section 20 of the 1985 Act; Miss Keen has raised a number of interpretation issues under the lease; and Mr Straw's correspondence could be taken to be an admission, but on the other hand it seems to raise complaints about management standards. In the circumstances he decides to allocate the case to the standard track.

Subsequently, Ms Reddy receives a list of the Tribunal's directions (largely confirming those mentioned at the PTR by Mr Hand) together with notice of a hearing in two months' time. The directions include the following:

- a requirement that each respondent provides a fully particularised statement of case
- a direction to the applicants to file supporting papers to the service charge accounts and replies to the respondents' statements
- an order that the applicants' representatives must prepare a fully indexed and paginated bundle of documents for the hearing (with copies for each party and each tribunal member), the contents of which are to be agreed with the respondents if possible
- a timetable for the hearing including an inspection of the property beforehand.

No direction for expert evidence has been included. The tribunal chairman was satisfied that no issue has yet arisen that requires an expert (and Ms Reddy is capable of acting as the applicants' expert anyway).

The LVT also sends an invoice for the hearing fee (£150), which must be paid within the prescribed time or the case will be treated as having been withdrawn.

The management company's strategy

Once the respondents' statements are to hand (or the date by which they should have been received has past), the parties and their representatives hold a meeting to take stock. Still nothing has been heard from Mr Munny, but Mr Chambers and Ms Potts of flat 12 have been in touch to say that they have no dispute and that they have asked their mortgage lender to settle their arrears on their behalf.

Mr Hickok and Miss Oakley have added nothing, but Miss Keen and Mr Straw have combined forces. They have instructed solicitors (Miss Keen's own firm, Eager & Co) who have produced a lengthy statement of case denying their clients' liability to pay the service charges on a long list of grounds, including:

- Tumbledown Management Ltd "TML" is in breach of section 20B

of the Landlord and Tenant Act 1985 by failing to notify expenditure within 18 months (see later in this chapter)
- the service charges generally are unreasonable and excessive
- insurance premiums are excessive
- the charges for external redecorations were unreasonable and works were carried out to a poor standard
- the management fees paid to Reddy & Willing are unreasonable and management services have not been provided to a reasonable standard (especially in the light of TML's failure to take adequate action against noisy residents)
- TML has failed to comply with section 21 of the 1985 Act as amended by section 152 of the Commonhold and Leasehold Reform Act 2002.

Eager & Co have also applied on behalf of their clients for an order under section 20C of the 1985 Act excluding the costs of the proceedings from the service charge.

Even though Ms Reddy is pretty sure that section 152 of the 2002 Act is not yet in force (in which respect she is correct), and even though those present suspect that the reply has been drawn up by Miss Keen who has just put in every point she can think of, the feeling of the strategy meeting is that the proceedings have now become sufficiently complicated to warrant bringing in TML's solicitors, Swift & Sharp, to take the case through to the hearing.

The hearing

The hearing takes place before a Tribunal consisting of Mr Hand (solicitor chairman), Ms Price (a valuation surveyor) and Ms Lyons (lay member). The applicants are represented by counsel (Mr Payne) instructed by Swift & Sharp, and Ms Reddy and TML directors Mark Bench and Walter Martinet are also present. Of the respondents, only Miss Keen and Mr Straw attend, represented by their solicitor Mr Eager.

The hearing takes most of the day allocated to it, with most of the time concentrated on Ms Reddy's examination and cross-examination and subsequent legal argument on the limitation points under section 20B. The tribunal chairman advises the parties that the decision is likely to be sent out to them in two to three weeks' time.

When the decision arrives, it is clear that the applicants have been successful pretty well throughout. Some minor adjustments need to be made to a cleaning account where charges could not be substantiated by documentary evidence but otherwise the tribunal has determined that the service charges claimed are payable. In the circumstances the tribunal has declined to make an order under section 20C excluding the applicants' costs from the service charge (although they have questioned as an aside whether the leases permit such charges in any event).

Within 14 days of the decision, Mr Straw pays his arrears and payment has also been received from the mortgagees for Mr Chambers and Miss Potts. Miss Keen having already paid, this leaves Mr Hickok and Miss Oakley (flat 3) and Mr Munny (flat 17) as the outstanding defaulters.

Meanwhile, Reston Laurels plc receives Swift & Sharp's bill for the LVT case: including counsel's fees, the bill is in the order of £2,500. The freeholder passes the bill to Reddy & Willing and suggests that it be added to the service charge if it cannot be recovered from the leaseholders concerned.

Enforcement

From the point of view of the service charge arrears, there is no longer a bar to the commencement of forfeiture proceedings in respect of flats 3 and 17. Service charges are reserved as rent in the Tumbledown Mansions leases, so it is not necessary to serve a notice under section 146 of the 1925 Act.

The cases will work out procedurally much as the scenario in chapter 5 regarding a forfeiture action for non-payment of rent. There will be some differences however:

- although all the cases could be contained in a single LVT case, two forfeiture actions (one for each flat) will be required
- the documents in support will need to include the LVT decision, and it should be referred to in the particulars of claim (at paragraphs 4(a) and/or (b) and 5)
- the particulars of claim (paragraph 4(b)) should refer the court to the covenant of which the leaseholders are in breach
- the claimant (which must be Reston Laurels plc) may be looking to TML for an indemnity as to its costs, and this is likely to involve discussion of how the costs of the LVT proceedings are to be claimed
- it is more likely than in a rent case that one of the leaseholders will attempt to mount a defence.

Perhaps most crucially from the practical point of view there is the aspect that both sets of leaseholders hold their flats subject to a mortgage. The agents' first action on receipt of the LVT decision is likely to have been that copies were sent to the mortgage lenders with a covering letter pointing out that the freeholder's way to forfeiture is now clear and inviting the lenders to intervene and make payment at this stage to save the additional costs of possession proceedings. It is probable that at least one of the lenders will have taken up that invitation.

In the event that any possession proceedings are necessary, the interest of the lenders must be noted at paragraph 9 of the particulars of claim so that the lender is served by the court.

If the case gets that far, the mortgagees are entitled to seek relief from forfeiture in their own right (see Chapter 2).

Costs (of the LVT proceedings)

The trickiest strategic problem still to overcome for the freeholder and the management company is the recovery of the costs so far incurred, especially the legal costs of the LVT proceedings, but the management time expended by Reddy & Willing will also have been considerable — and it is likely to be outside the terms of their agency agreement. Reddy & Willing's bill to TML could easily exceed £1,000, leaving around £3,500 to be recovered somehow.

The freeholder will have ensured that it does not have to contribute from its own funds to the cost of recovering service charges in which it has no interest. The management company meanwhile has no assets of its own and no income save for the service charge or what it can bring in by way of a levy on its members. TML's policy is not to penalise leaseholders who have paid their service charges promptly by charging them to recover arrears from defaulters, so what are their options?

- TML may take the view that the costs of the LVT proceedings should be divided equally between the leaseholders of the five flats whose arrears sparked off the decision to go to the LVT, but they accepted payment at an early stage from Miss Keen and especially from the mortgagees of Mr Chambers and Miss Potts without any indication of the scale of costs which they might be seeking to recover later. It is probable that the mortgagee will decline to pay.

- Perhaps it would be more reasonable to charge the costs to the remaining defaulters and those who contested the proceedings, leading to the decision to instruct lawyers, but how to do the sums most fairly? Also, opposition could be expected (at least from Miss Keen), and there are other problems as well.
- Taking the line of least resistance, TML could split the costs simply between the two leaseholders who were still in arrears by the time forfeiture becomes permissible, but they will be left in difficulty if the mortgagees refuse to pay more than a nominal amount — as is predictable.
- Whichever of the above is attempted, the costs will only be enforceable properly if the lease allows it. For example, a covenant to pay costs "of and incidental to the preparation and service of a notice under section 146 of the Law of Property Act 1925" (a typical form of words) would be inadequate for two reasons: no such notice can be served until the LVT proceedings are concluded so the clock has not started to run; and no notice could be served in these cases anyway because the service charges are reserved by the lease as additional rent.
- Even if costs are recoverable from the individual leaseholders in default, they will be subject to reasonableness and open to challenge and subsequent scrutiny by the LVT as administration charges (see Chapter 13 and elsewhere).
- If the costs cannot be put to the individual leaseholders then TML must look to the service charge; however, they are only recoverable through the service charge in so far as the lease allows it and the tendency is to construe such provisions restrictively. Indeed, in our scenario, the Tribunal has already cast doubt on this point.
- There is one final hurdle to be overcome, even if TML has found a way through the preceding points: and that is the question of the legality of recovering the costs of LVT proceedings at all. Paragraph 10(4) of schedule 12 to the Commonhold and Leasehold Reform Act 2002 provides that no-one can be required to pay another's costs incurred in LVT proceedings except by determination by the LVT within its limited powers. This would seem to suggest that contractual provisions for costs (such as a covenant in a lease covering the costs of enforcement or forfeiture) will be of no effect in so far as costs at the LVT are concerned. Unfortunately there is no judicial authority to elucidate this point at the time of writing.

The above list of options would have been fore-shortened if the LVT had already made a determination on the whole question of the parties' costs of the proceedings. Unlike in court proceedings, the question of an award of costs is not raised on the face of an LVT application. None the less, it is open to a party to invite the tribunal to use its powers to award costs. (The LVT's powers in this respect are described in Chapter 3 and discussed in Chapter 13.)

Mr Payne, as the applicants' counsel, getting wind of the drift in his clients' favour, could well have submitted to the tribunal that they should use their powers to award his clients their costs (up to the statutory maximum) on the basis that they had been successful and that the respondents' case (the effect of which had been to increase drastically the applicants' costs) had been manifestly without merit or any real prospect of success — to the extent that the respondents had acted unreasonably. Advocates brought up in litigation in the High Court or county court would find such an application only natural.

In the LVT in circumstances like these however, advocates must be cautious before requesting costs awards from the tribunal. Whatever the outcome, problems may be created for their clients:

- if the tribunal took the view that the respondents' conduct was not such that they had acted unreasonably (the likeliest result), the chances of Reston Laurels and TML subsequently claiming the same costs in any other way would be significantly diminished
- alternatively, if costs were awarded they could be no more than the statutory limit of £500, leaving a shortfall of £3,000 which might be almost impossible to recover — a court being likely to decide that the applicants should not have two bites of the same cherry.

Overall, it will probably be preferable for the freeholder and management company simply not to ask the LVT for an award and seek to rely instead on the costs covenants in the leases.

The strategy on costs needs to be decided at an early stage, certainly before the hearing. The variety of potential outcomes should be considered and weighed in the light of the particular case, including the provisions of the leases and the amount of the costs. For example, in a straightforward LVT case where the advocacy was conducted purely by the managing agents, the LVT's maximum allowable award might be perfectly adequate.

As always, the costs issue is likely to turn out to be the predominant tactical concern of the parties to any litigation, and LVT proceedings will be no different. In our scenario, even with relatively little recourse to lawyers, the action could have proved to be uneconomic for a management company with no independent resources. Clearly property managers who can offer their clients cost-effective arrears recovery will have an increasing advantage in the market.

Forfeiture via the county court (without reference to the LVT)

It is not compulsory to apply to the LVT to determine that a breach of covenant to pay service charges has occurred. Section 170 of the Commonhold and Leasehold Reform Act 2002 amends section 81 of the Housing Act 1996, but not that drastically; the determination can still be made by the court (or indeed under a post-dispute arbitration agreement).

This does not mean that a landlord can go straight to the county court with a possession claim and ask within those proceedings for a determination of the breach. That is ruled out (in service charge cases) by section 81(2) of the 1996 Act, and it would fly in the face of the intention of the 2002 Act.

Landlords may still apply to the county court for a declaration that a leaseholder is in breach of covenant, and then use that declaration as the basis for a forfeiture action, even in a service charge case. The perceived advantage of applying to the court is that, if successful, the landlord has a much greater chance of recovering a significant proportion of his costs.

Particularly in service charge cases however, there is a strong probability that the matter will be transferred to the LVT. This is true for a number of reasons including the public policy implications, the natural inclination of judges to avoid service charge issues, and the drive towards reducing the costs of litigation.

If a case is transferred there will be obvious additional delay, but of more serious concern to landlords will be that the court could take the view that the case should never have begun there and that the landlord should thus lose his costs so far and pay any incurred by the defendant.

In all the circumstances, unless there are other genuine breaches of covenant upon which the landlord can rely (in which case the court is more likely to keep the case), it is probably prudent to issue most claims for service charge determinations in the LVT.

In the event that the case stays in court, for whatever reason, the procedures to be followed in a claim for a declaration are considered in more detail in Chapter 8.

Having obtained a declaration, the landlord may proceed to serve a section 146 notice if the service charges are not reserved by the lease as additional rent (see Chapter 2). (A scenario with a precedent section 146 notice can be found at Chapter 9.) In any event it is standard practice to invite any mortgage lenders to intervene at this stage. There are distinct advantages to the court approach in this context: first, lenders are much more likely to pay the costs of court proceedings without question; second, a declaration of the court involving the leaseholders of a single flat is considerably easier to understand from the face of the document than a decision of an LVT which is not set up as a judgment and may cover a large number of respondents.

If matters have still not been settled after declaratory proceedings and a section 146 notice then the landlord may have to forfeit the lease by the issue of possession proceedings. Procedurally these will follow the example in Chapter 5, but see also the comments under the heading "Enforcement" earlier in this chapter.

Overall, the principal advantage to proceeding through the court is often seen as the greater prospect of recovering more in costs. That may be a false perception however. The risks include:

- transfer to the LVT and consequential costs penalties (as well as additional delays in starting again)
- a shortfall in costs recovery caused by the courts' constant pressure on matters such as rates of charge and proportionality
- the overall bill may be so much higher that even the recovery of a greater proportion does not overcome a larger shortfall.

Debt recovery

Most of the complications referred to in the earlier parts of this chapter only arise if forfeiture is contemplated or threatened. The complications derive largely from the various obstacles devised by Parliament and the courts to restrain the worst excesses of forfeiture. No such restraints are put in the way of simple actions for debt which involve no threat to leaseholders' homes and from which there is no risk of disproportionate windfall profits for landlords who are seen as unscrupulous.

The details of a debt recovery procedure are examined in Chapter 7, but it should not be overlooked that there is still a slight risk of a debt case being transferred to the LVT as the tribunal's jurisdiction is not restricted to forfeiture cases alone. The possibility of transfer only arises in the event of a disputed case which comes before a judge. In a debt case (unlike in a possession action) judgment can be entered in default of a defence without the need for a hearing.

Even though a default judgment cannot properly form the basis for forfeiture and despite the fact that there will have been no judicial determination of the case, some lenders will accept it as sufficient to enable them to discharge arrears on their borrowers' behalf. This practice has diminished somewhat since the implementation of the Commonhold and Leasehold Reform Act 2002, but not as much as expected; it depends upon the policy of the lender concerned and the wording of their mortgage deeds. Even for lenders who currently accept county court judgments in this way, there is no guarantee that they will continue to do so, so there is a possibility that landlords who attempt this strategy could be left high and dry.

Limitation

Limitation of actions is one of those tricky legalistic areas which cause occasional hiccoughs in litigation because the rules can seem unnecessarily complicated. The central point is that a claim will be time-barred (or "statute barred") if it is of a class covered by the Limitation Act 1980 and the cause of action has existed longer than the limitation period allotted by the Act to that class of case. The best known examples are that a claim under a simple contract has a limitation period of six years, but a claim under a deed under seal has 12 years.

Relevant periods under the Limitation Act for leasehold cases include:

- actions on a specialty (such as a lease made under seal) 12 years (section 8)
- sums recoverable by statute (which may be relevant in service charge or enfranchisement cases — see below) six years (section 9)
- recovery of land (such as by forfeiture) 12 years (section 15)
- recovery of rent (whether by debt recovery or distress) six years (section 19)
- actions by beneficiaries under a trust (such as leaseholders in respect of service charge funds) no limitation (section 21).

As can be seen, the minimum period for relevant limitations is six years. Any reader with a claim which exceeds six years in age is advised to take specific legal advice on the status of their claim.

However, limitation periods only apply to cases to be brought before a "court of law". What is uncertain is whether an LVT falls within this description. If it is not, then there is no statutory limitation period prohibiting particular cases from being brought before it (although any follow-up applications to the courts for enforcement and so forth will be caught).

There is an interesting discussion of this problem and limitation generally in the LVT case of *Murphy* v *St Andrew's Square (West) Management Co Ltd* (LEASE Ref: 1143) where the tribunal in that case took the view that an LVT is a "court of law", following a Court of Appeal decision that the Lands Tribunal is such a court. The decision is neither conclusive nor a precedent however and it seems likely that the point will be re-visited before an appellate court in due course.

Limitations on claiming service charges

Property managers must also bear in mind that they are subject to an additional time constraint when it comes to service charges: that contained in section 20B of the Landlord and Tenant Act 1985.

Section 20B provides that service charges cannot be recovered if the leaseholder has not been given written notification within 18 months of the cost being incurred that he will be required to contribute. The written notification need not be a demand, but a demand will obviously suffice. This is known as "the 18 month rule" and it can cause major problems for the unwary.

Agreeing or determining service charges prior to expenditure

As a final thought in the context of service charge recovery, the best plan is to avoid having to take recovery measures by maximising reserves and cash flow prior to expenditure.

Generally, it is preferable to agree budgets and payment plans with leaseholders, especially when it comes to substantial expenditure. This can be done alongside consultation under section 20 of the 1985 Act, but it is important not to skimp on the consultation procedures even if the expenditure is agreed in advance. While enforcement is subject to what is allowed by the lease, there is nothing to prevent leaseholders from agreeing voluntarily to go beyond their obligations in the lease; after all, efficient and cost-effective management should be in the interests of all concerned.

An alternative method is to apply to the LVT, before embarking upon major expenditure, for a determination that the proposed costs are reasonable. While this can give the property manager some reassurance, it is not foolproof. Leaseholders are entitled to go to the LVT after the works to argue that it was not done to a reasonable standard or if the final cost exceeds the estimates. Furthermore, a preliminary determination of the reasonableness of the cost will not protect a manager who fails to follow the section 20 consultation procedures correctly. A defective consultation process will prevent the landlord recovering more than the consultation thresholds however reasonable the cost.

None the less, in general terms it is correct to say that the best way to recover service charges is to take all reasonable steps to ensure from the outset that there is no room for dispute by maximising cooperation with leaseholders and minimising procedural errors.

7 Debt Recovery in the County Court

It is now extremely unusual for straightforward litigation concerning residential leaseholds to take place in the High Court, whether it be in terms of debt recovery, contractual disputes or possession actions. Occasionally, cases will be transferred to the High Court for enforcement after judgment, but that is also less frequent than it was. Consequently, this chapter will focus exclusively on the county court; in any event, there is little difference procedurally under the Civil Procedure Rules "CPR".

Procedures

Pre-action

The first step in pretty well any litigation these days is not part of the proceedings as such, but is required by the rules of court: namely that the claimant should comply with the pre-action protocols. There is not (as yet) a specific protocol in force for debt recovery cases but the letter before action should comply generally with the requirements of the Practice Direction on Protocols under the Civil Procedure Rules with respect to letters of claim. Essentially, it should give a clear and concise summary of the facts and what is sought. Reference should be made to the availability of independent legal advice. The letter should give a reasonable time for a reply (perhaps 21 days) and inform the recipient that proceedings will be commenced without further notice if no satisfactory reply is received by a given date. If any information is requested, a reasonable time should be allowed for any necessary investigation and reply.

The purpose of the letter is principally to inform and it should not be confrontational or intimidating. Underlying that is the idea that its delivery may prompt an offer of settlement. Letters which do not comply with that ethos are likely to lead to criticism from the court.

The protocols also require the defendant to take certain steps. The defendant should acknowledge the letter of claim and either state that he accepts the claim (with settlement proposals) or that it is disputed and, if so, why. If a dispute exists, the parties are expected to make all reasonable attempts to reach a negotiated settlement before resorting to proceedings.

Failure to comply with the pre-action protocols may be penalised by costs orders or even dismissal in extreme cases.

Generally, the rules of court change from time to time. It is advisable to keep up to date with the information which can be obtained easily from any county court office or the courts service website *www.hmcourts-service.gov.uk*.

What follows is a case study looking at a fairly realistic situation where a management company elects to collect a service charge debt through the county court rather than attempting forfeiture.

Scenario 3: Tumbledown Management Limited v Andrews

TML, although not the freeholder of Tumbledown Mansions, is the manager party to the leases and so is entitled to the benefit of the leaseholders' covenants to pay service charges. Consequently it is the party which has the right to sue for service charge arrears (although it is not entitled to take forfeiture proceedings — see Chapter 2).

The Andrews sisters, Holly and Ivy, are leaseholders of flat 18, Tumbledown Mansions. They are in arrears with their service charges in the sum of £2,000. They have a mortgage over their flat. The sisters are both self-employed as partners in their own firm of interior designers.

In keeping with their usual practice in debt claims, TML authorises its managing agents Reddy & Willing to take proceedings in the local county court for recovery of the arrears. Reddy & Willing have sent the sisters a letter before action (which is a standard letter prepared in consultation with their solicitors to check compliance with pre-action protocols) but no response has been received after 21 days so Ms

Reddy asks a member of staff to prepare the papers for the county court. (Reddy & Willing continue to issue debt claims physically and by post. Readers are advised to consider whether issuing electronically is better suited for their purposes: see *www.hmcourts-service.gov.uk* and click on "Money Claim Online".)

Issuing proceedings

Ms Reddy's assistant, Ms Starr AIRPM, prepares a claim form on county court form N1, based on a precedent created in consultation with the firm's solicitors. Ms Reddy checks and arranges for it to be signed by TML's chairman, Mr Martinet, drawing his attention to the fact that he is signing a statement of truth and the import of that. It is vital that the individual signing the form has the proper authority to do so (a managing agent is not an appropriate person) and satisfies himself that the contents of the claim form and particulars of claim are accurate. The statement of truth verifies the accuracy of the remainder of the document and the maker of a false statement can be punished for contempt of court (the penalties for which are severe).

This claim form is reproduced in Appendix 4.

The claim form and its attachments (plus two copies of each for service upon the defendants) and a cheque for the court fee are delivered to the court office. The court clerks then issue the proceedings and serve the papers on the defendants by post, together with a "Response Pack" which consists of an admission form (N9A), a defence and counterclaim form (N9B) and a form for acknowledging service of the claim. The response pack explains to the defendant what needs to be done in a few well-chosen words. In particular, it explains that a defendant who disputes the claim may effectively gain a further 14 days by filing an acknowledgement of service indicating either an intention to defend all or part of the claim or a challenge to the court's jurisdiction.

To take the last point first, it is possible that a defendant may challenge jurisdiction in a service charge case on the basis that it should go to the LVT. On its own such an application should not succeed because the county court has jurisdiction concurrent with that of the LVT, but the case may still be transferred if the evidence discloses a substantial dispute on the reasonableness of the charges claimed. It would be very unusual however for the case to be transferred at such an early stage.

To summarise the position, the range of possibilities depending upon the action taken by the defendants is:

- no action of any sort — the claimant may enter judgment in default fourteen days after service of the claim form
- the defendants file an admission — the claimant needs to consider any offers of payment accompanying the admission or look at the alternative methods of enforcement (see below)
- a defence is filed (with or without counterclaim) — the case will proceed
- an acknowledgement of service is filed but not followed up within 28 days of service of the claim form — the claimant may enter judgment in default.
- an acknowledgement of service is filed and followed up by a defence — the case will proceed.

(In our case, the Andrews sisters file a defence, but see below for general notes on default judgments and admissions.)

The defence and allocation

The defence must be filed either within 14 days of service of the claim form or (if an acknowledgement of service was filed) within 28 days; otherwise the claimant is at liberty to enter judgment in default. Judgment can be set aside later, but it is not guaranteed and considerable inconvenience and expense is caused all round in any event. The form supplied with the response pack can be used, but it is not essential — solicitors will generally create a word processor document for example — but the basic information on the form should be present (eg case number and parties, details of the defence, a signature to a statement of truth, and the defendants' address for service).

The Andrews sisters have been to see a solicitor who has helped them to draft their defence (see Appendix 4). The defence arrives at Reddy & Willing's office together with an allocation questionnaire (N152). Both sides receive an allocation questionnaire, the aim of which is to assist the court in deciding to which track the case should be allocated. A helpful leaflet entitled *The defendant disputes all or part of my claim* (EX306) published by HM Courts Service deals simply with the question of allocation. (It is common among many courts not to trouble the parties with allocation questionnaires in money claims for less than £1,500 as these will almost invariably be allocated to the small claims track.)

Ms Starr completes the allocation questionnaire in discussion with Ms Reddy (the form appears at Appendix 4). In looking at the form and considering the defence, Ms Starr (who has attended a course on county court litigation) is prompted to suggest that an application for summary judgment would be appropriate in this case. Summary judgment may be awarded if the court accepts (to paraphrase part 24 of the CPR) that the defendants have no real prospect of successfully defending the claim and that there is no other compelling reason why disposal of the claim should await a full trial.

Other useful applications which could be considered at this stage of the proceedings (and which may confer a tactical advantage) include the striking out of an opponent's statement of case (the particulars of claim, defence or counterclaim) and requests for further information (formerly known as "Further and better particulars").

The court may strike out a statement of case (under CPR 3.4(2) if:

- it shows no reasonable grounds for bringing or defending a claim
- it is an abuse of the court's process or is likely to obstruct a just disposal of the proceedings
- there has been a failure to comply with a rule, practice direction or court order.

In practice, the power is rarely used if there is any alternative way of curing the defect in the party's case. To give in too easily to the temptation to apply for a strike-out may make the applicant liable for penalties.

Requests for further information are slightly different because they may be made direct to the other side without referral to the court. The court may need to become involved for enforcement or for protecting a party from unduly oppressive requests. Requests should be concise and confined to points which are "reasonably necessary and proportionate to enable the first party to prepare his own case or understand the case he has to meet" (PD 18 para 1.2 of the CPR).

Once again in practice the court is unlikely to be particularly draconian regarding the statements of case (formerly "pleadings") of unrepresented parties in small claims. Of the three types of interim application mentioned (summary judgment, strike-out and further information), summary judgment is the most likely to be used in a small claim; even then, it is not used widely by unrepresented parties. Guidance on the variety of applications available can be obtained from a range of sources, including county court offices and the courts service website. Each of these types of application can be made against any party (including the claimant).

The summary judgment application

Ms Reddy and Ms Starr discuss the question of the proposed application at a convenient TML board meeting and obtain authority to proceed with it on the basis that the defence is effectively superseded by the LVT decision (reported in Chapter 7 as scenario 2) which had arrived recently. Even if that decision had not been made by the LVT, an application for summary judgment may still have been worthwhile as the defence makes no real assertions challenging the amounts claimed. The application (which is on county court form N244) can be found at Appendix 4.

Fees are payable both on the filing of the allocation questionnaire and on the application for summary judgment. Ms Starr files both documents together (having previously sent a copy of the allocation questionnaire to the Andrews sisters, inviting them to agree to its contents). The court clerk places the completed allocation questionnaire on the court file and arranges a hearing before the district judge, sending notification of the date and time and details of the application to the parties.

The summary judgment hearing

Ms Reddy attends the hearing to represent TML, but (from previous experience) she is aware that the district judge will allow the matter to proceed on that basis only if a director of TML is present, so Walter Martinet is also present. Both the Andrews sisters attend.

The case is heard by District Judge Scales who, having explained briefly the nature of the hearing and invited the parties to outline their positions, concentrates on the LVT decision and the proceedings which led to it. The judge notes that the Andrews sisters were not parties to those proceedings. Although they had been served with copies and had chosen not to apply to be joined, the judge feels that it would be unjust to regard them as bound by the LVT's findings. She also notes however that the sisters have not put forward any cogent arguments supporting their allegations that the charges are unreasonable and that they have not specifically applied for a transfer to the LVT.

In the end, the judge takes the view that some points raised by the sisters at the hearing regarding the standard of the decorative works are sufficient to disclose a substantive argument justifying a full hearing. Given the history of the previous LVT proceedings and that this is a money claim rather than a forfeiture case, the judge decides not to transfer the matter to the LVT. She also feels that the defendants have been unable to demonstrate that they have a full defence to the claim and so she decides to impose a condition that the action should proceed to trial only on terms that the defendants pay half the amount of the claim on account within 14 days.

The judge then orders as follows:

1. the defendants shall pay into court within 14 days the sum of £1,000 to abide the event
2. if the defendants fail to comply with paragraph 1 of this order their statement of case be struck out and the claimant be at liberty to enter judgment
3. subject to compliance with paragraph 1 the following directions shall apply:
   (a) the claim is allocated to the small claims track
   (b) the hearing of the claim will take place at [time] on [date] at Backwater County Court, [address] and should take no longer than 90 minutes. *The court must be informed immediately if the case is settled by agreement before the hearing date.*
   (c) each party shall deliver to every other party and to the court office copies of all documents (including any witnesses' statements of fact/experts' report) on which he intends to rely at the hearing no later than 14 days before the hearing.

   The original documents shall be brought to the hearing.

The judge considered a number of other directions, including an order requiring the defendants to clarify their case and give details of their allegations concerning the decorations; however, possibly feeling that the task would be beyond them, she contented herself by noting down their verbal allegations at the hearing. The judge also considered providing for an expert's report, but decided against on the grounds of proportionality. She did warn the defendants however that, having failed to argue for expert evidence, it would be up to them to substantiate their allegations on the facts, and pointed out that she was ordering the exchange of witness statements.

Compliance with the court's directions

The court's directions are its various orders and requirements to be carried out before the hearing. They most commonly deal with disclosure of documents, exchange of factual and expert evidence, preparation of bundles of documents for use at trial and the arrangements for the hearing itself. Some directions can be backed up by specific punitive sanctions in the event of non-compliance. These are widely known as "unless orders" and they usually provide for a party's case to be struck out if there has not been compliance by a set date or within a fixed period. The sanctions are generally imposed after at least one failed attempt or missed deadline. Applications for "unless orders" are a useful tactical weapon.

It is important to comply with directions and not only because of the risk of "unless orders". If a party attends a hearing without having complied fully, they may be prevented from putting part of their case or the hearing may have to be adjourned, causing further delay and expense. A party in default may be ordered to pay the other side's "wasted costs" even in a small claim.

In this case, the only direction which concerns TML is that requiring copy documents to be supplied to the defendants and the court, with arrangements made to take the originals to the hearing. The documents here will include witness statements, but not experts' reports. No doubt Reddy & Willing will wait first to check that the Andrews sisters have complied with the order to pay into court. This can be done by a telephone call to the court.

Documentation

Other than the witness statements, the documentation needed in a case such as this comprises:

- the lease (to prove the claimant's contractual entitlements and the Defendants' obligations)
- office copy entries from HM Land Registry (to establish that the claimant and defendants are the appropriate current parties)
- a statement of account showing the arrears claimed
- correspondence relating to any items in dispute
- year end accounts for the periods in question
- receipted invoices for expenditure in dispute.

Of course, the first three items should be covered by what has already been provided as attachments to the claim form and particulars of claim.

It is best not to include too much documentation, but common-sense should be applied selectively to the files of TML and Reddy & Willing to ensure that any paperwork which might shed light on the case is produced. For example, there may be something in the section 20 consultation documents and the estimates which deals with specific points raised by the defendants. Also, given the point in the defence concerning management fees, it would be sensible to provide the management agreement between TML and Reddy & Willing.

If a large volume of a documents is to be provided (which is almost inevitable in a leasehold case) it should be contained ideally in an indexed and paginated bundle. Solicitors will be expected to do this, and it is probable that the court will expect professional managing agents to do it as well. In any event, thoughtful preparation will go down well with the judge at the hearing.

The parties' duties relating to disclosure of documents are extensive. They are set out in part 31 of the CPR and its associated practice direction. The formalities affecting disclosure in cases on the small claims track are less onerous than in fast track and multi-track actions, but the general principles are essentially the same.

- "Documents" means anything upon which information is recorded, so this could include photographs, computer files and disks, emails or tapes as well as paper.
- The parties are required to disclose documents in their control which are material to the issues in the claim, notwithstanding that they may adversely affect their own case or support another party's case.
- Documents within a party's control cover those which are or have been in that party's possession or over which they have been entitled to exert rights of possession, inspection or copying (such as documents in the possession of a managing agent).
- Parties must make a reasonable but proportionate search for relevant documents. What is "reasonable" or "proportionate" will obviously vary, but parties may have to justify to the court their decisions not to search.

The parties may be ordered to give disclosure by serving a list of documents. There is a prescribed form for this purpose (county court form N265) and care should be taken to follow the guidance notes on the form.

Clearly it will not always be realistically possible to take originals to the hearing, but the court places great importance on inspecting original documents whenever possible, so it is in the parties' interests to make every effort in this regard.

Witness statements

Many parties in small claims overlook the requirement to file witness statements as to fact prior to the hearing, possibly believing that their presence at the hearing will suffice. They will not always be let off, and this applies particularly to those who are professionally represented. If a statement has not been served for a particular witness before the hearing, the court may refuse to hear oral evidence from that witness. Conversely, just because a statement has been served does not mean that that witness's evidence will inevitably be allowed even if he does not attend the hearing: unless otherwise directed, statements will only be admitted into evidence if the maker is available for examination. In any event it is good practice to assemble witness statements because it makes the preparation and conduct of the case at the final hearing so much easier.

In this case the statements should be from the witnesses identified in the allocation questionnaire: Ms Reddy and Mr Martinet.

The formalities required for witness statements are set out in part 32 of the CPR. An example of the "top and tail" of a statement can be found in Appendix 4, together with notes on the requirements for statements and exhibits.

What should be remembered at each stage of the litigation process is that a claim (or a defence) has the best chance of success if it is clearly set out and fully supported by evidence and documentation. The court's directions are intended to assist the parties put their cases most effectively — not simply to cause inconvenience and expense. The court cannot decide a case unless it has the full facts and an adequate explanation of the parties' positions.

The final hearing

For the purposes of this illustration, we shall assume that the directions have been complied with and that all the parties and witnesses who should have attended are present for the final hearing, thus avoiding the need for any adjournments.

The hearing is again before District Judge Scales, who should be addressed as "Ma'am" (a male district judge would be "Sir"). Particularly as there are no lawyers present, the judge explains the nature of the hearing and how it is to be conducted. The hearing is relatively informal in that the parties remain seated and evidence is not taken on oath; however, deference is shown to the judge and the parties should behave in a courteous and civil manner. The hearing takes place in the district judge's hearing room rather than the main court-room and no special formality of attire is required, although smart dress is expected (especially from the advocates).

Ms Reddy asks for permission to act as lay representative (that is, as an advocate who is not legally qualified) for the claimant, and the judge accedes to her request.

The judge then invites Ms Reddy briefly to introduce the claim, explaining its essential features and pointing out which areas are disputed. Holly Andrews (as spokesperson for the defendants) simply agrees that Ms Reddy's outline is accurate. Ms Reddy then refers the Judge to the claimant's witness statements. The judge says that she has read them and that she is happy to let them stand as their evidence in chief. Ms Reddy has nothing to add to the statements.

The judge then asks if Ms Andrews wishes to cross examine the claimant's witnesses.

The next period of the hearing is more argumentative, with the defendants simultaneously putting questions and submitting their own points about the standard of decorative works, and one or two issues which they had not raised earlier. The judge allows the defendants some latitude, but eventually brings the hearing back to focus on the main issues in the case.

The evidence having concluded, the judge thanks the parties for the way in which they have prepared their cases, and proceeds to deliver her judgment. The judgment is in favour of the claimant. The judge concludes by telling the defendants the limited grounds for appealing her decision, but they do not indicate any intention of doing so.

Ms Reddy has calculated the interest up to the date of the hearing, and the judge allows that to be added to the judgment. The judge also allows TML to recover the fees paid to the court (the issue fee and the fees paid on allocation and for the application for summary judgment) and asks Ms Reddy if there is anything else. Ms Reddy asks for the Land Registry fees for obtaining office copy entries, and they are also allowed.

Feeling things are going so well, Ms Reddy asks if TML can have an order for its costs in relation to her own fees. She calculates that she has incurred at least thirty hours of time between herself and Ms Starr. She puts this claim at £1,500. The judge asks if the defendants had been warned of this claim and, in particular, if TML (or Reddy & Willing on their behalf) had served a costs statement before the hearing. Ms Reddy confirms that they had not done so.

The judge dismisses the claim for time costs, first, because a costs statement had not been served (see under "Assessment of costs" below) but second, because this is a small claim and she has not been persuaded that the defendants' conduct was such that costs should be awarded outside court fees and disbursements. (The question of costs is discussed more generally below.)

All that is left is for the defendants to discharge the judgment within 14 days as ordered or, failing that, for TML to enforce it. From this point on, TML is in the same position it would have occupied had judgment in default been possible.

Default judgments

As we have already seen, a default judgment is possible in one of three ways:

- no response is made to the claim within 14 days after service of the claim form
- an acknowledgement of service is filed but the defendants take no further steps, all within 28 days of service of the claim
- the defendants file an admission of the claim.

With either of the first two eventualities, the claimant is free to enter judgment by completing the appropriate form (county court form N225, available from county court offices or *www.hmcourts-service.gov.uk*) and then to proceed to enforce its judgment. The position is different however if an admission is made which incorporates an offer of payment.

Offers to pay

If a defendant makes an offer of payment by instalments, the claimant can accept the offer (in which case the court will issue an order requiring the defendant to pay in accordance with his offer) or object to it. If the claimant objects, he can state what payment terms would be acceptable (usually payment forthwith). The payment terms which are eventually fixed are decided by the court. The claimant is entitled to make representations at a "disposal hearing" when the court will enter judgment and decide on payment terms.

There are some strong arguments which landlords and management companies can put forward to resist long-drawn out instalment schemes; for example:

- payments will not keep up with current service charge instalments, so the defendant's account will never be up to date
- the admission now having been made, the defendant's mortgage lender will be willing to pay in full (see below)
- (in the case of resident-owned management companies) allowing the defendant an inordinate time to pay would be manifestly unfair to other leaseholders who would be effectively subsidising the debtor
- non-payment is holding up essential works of repair or maintenance, causing risk of personal injury or damage to property
- other methods of enforcement are available which will realise the debt more swiftly.

The final decision rests with the court however, and there is a risk therefore that the landlord or management company will have gone through debt recovery procedures to achieve an instalment order which sees payment coming through no more quickly than if they had simply waited for the defaulting leaseholder to sell his flat. In order to minimise this risk, managers should grasp the opportunity to object to instalments and have thoroughly prepared arguments ready.

Administration orders

Another situation may arise whereby the judgment creditor may have to take decisive steps to avoid being caught by long-term instalments or even under recovery. The court, a creditor, or more usually the debtor may initiate the process by which the court makes an administration order which, although nothing directly to do with the insolvency legislation, is effectively a step short of bankruptcy to administer the subject's debts.

An administration order commonly provides for the debtor to make periodic payments to creditors specified in the order over a period of three years at rates fixed by the court. The provision may not cover full payment of the debts. It is possible for a creditor to opt out of the arrangement, and it is generally wise to do so — especially if there are better prospects of recovery by other means. (Creditors are given notice by the court and an opportunity to object to their inclusion.) However, if the creditor is included in the scheme, it will not be possible to use other enforcement measures while the administration order is in force.

The position of mortgagees

As discussed briefly in Chapter 6, some mortgage lenders continue to regard a debt judgment as sufficient determination of a service charge claim to enable it to discharge its borrowers' arrears without further reference to the borrowers, even though the judgment may have been obtained without any determination by a judge. This practice varies between lenders, and even between regions of the same lender, and it can hardly be relied upon. None the less, a landlord with a judgment debt in his favour is almost bound to invite the lender to intervene to save the cost and delay of further enforcement measures.

Enforcing county court judgments

Once judgment has been obtained, the successful claimant becomes the "judgment creditor" and is able to enforce his judgment through any of the prescribed methods. More than one method may be used, either at the same time or one after another (rule 70.2 of the Civil Procedure Rules 1998) — but obviously not in order to recover more than the sums contained in the judgment debt.

The principal methods of enforcing a county court judgment are:

- warrants of execution (a visit by the bailiff)
- third party debt orders (formerly "garnishee orders")
- charging orders (charging the debtor's property, such as the flat).
- attachment of earnings orders
- orders to obtain information from judgment debtors (formerly oral examinations). (This is not a method of enforcement *per se*, but an examination may be necessary to acquire sufficient information of the debtor's means. Moreover, the process itself can be so embarrassing or intimidating that the debtor becomes motivated to pay up. Also, if the debtor fails to respond to an order quickly enough he/she becomes liable potentially for contempt of court, which might conceivably lead to imprisonment.)

Each of these enforcement measures is discussed in Appendix 4. It is worth remembering that interest continues to run on judgments for over £5,000, and this should be calculated and added whenever possible to enforcement applications.

The enforcement options are very much "horses for courses". It is important to select the method which is most appropriate for the individual case, which will usually be determined by the debtor's personal circumstances. Property managers with large portfolios may too easily fall into the habit of favouring a particular approach across the board. Policies need to be reviewed from time to time if the best results are to be achieved for clients and cash flow.

Bankruptcy

An alternative to the standard methods of enforcement in the county court is to petition for the debtor's bankruptcy (or winding-up in the case of a company). This can be attempted as a substitute to obtaining a county court judgment or after judgment has been given. The basic ground for a petition is that the debtor is insolvent (that is, he is unable to pay his debts).

The alternative routes to a petition are:

- an outstanding judgment debt where a warrant of execution (or its High Court equivalent) was returned unsatisfied
- the prior service of a statutory demand (a prescribed form under the Insolvency Rules 1986) which the debtor has failed to comply with.

In either case, the debt must be at least £750.

A statutory demand should not be used if there is a substantial dispute, or the debtor has a set-off or counterclaim which at least equals the size of the debt, or if the creditor holds some security which equals or exceeds the debt. The recipient is entitled to apply to the court to have the statutory demand set aside in such circumstances, and the party who issued it inappropriately may be penalised in costs.

In practice, bankruptcy petitions are very rare in leasehold cases. There may be a number of reasons for this, such as:

- residents' management companies are understandably reluctant to make their neighbours bankrupt
- bankruptcy proceedings are very expensive and the prospects of recovering the full debt and the costs are not good
- leasehold situations are rarely free of dispute
- a bankrupt leaseholder may be an ongoing liability for some time
- the lease is generally thought of as better security in the long term.

Those considering bankruptcy as an option in debt recovery should take independent legal advice before making any policy decisions.

Costs

Costs allowed on entering a default judgment are on a fixed scale. These are known as "fixed costs" and cover the court fee and the relatively small amount shown on the claim form and entering judgment. The fixed costs system is rigid and it only covers solicitors' costs.

In fully defended cases the costs allowed to the victor will usually be appropriate for the type of case with multi-track the most generous, small claims the lowest and fast track in the middle. This will not always be the case however. First, the court is required to consider proportionality. Further, there will be other factors affecting the court's view. In any event, costs are always in the court's discretion.

A point often overlooked is that the court may award witness expenses even in the small claims track, subject to proof. This includes loss of earnings, although there is a limit currently of £50 per witness (who could be the party), or expert's fees of up to a total of £200 to include his report and attendance at the hearing (but it is rare for the court to give permission for expert evidence in a small claim). The power arises under CPR 27.14.

Other than expenses and fixed costs, the court's power to award costs in small claims arises only if the other party "has behaved unreasonably" (CPR 27.14(2)). There is no clear guidance as to what constitutes unreasonable behaviour, but the following have been regarded in that light in the particular circumstances of the case involved:

- late discontinuance of the claim or defence
- late admission of liability
- pursuing a claim or defence with no reasonable prospect of success
- unreasonable failure to negotiate
- unnecessarily delayed attempts to settle
- failure to state the party's case with adequate clarity
- presenting evidence in a misleading way
- dishonesty
- delay.

These are only examples however. Although the successful party in a small claim should consider the possibility of recovering costs over and above fixed costs, it must be remembered that such orders are still very rare and it is unwise to antagonise the court by requesting costs orders when there is no likelihood of obtaining them.

Assessment of costs

When costs are awarded and the successful party is legally represented, the court will assess the amount of costs to be allowed (unless the figures are agreed by negotiation) and the basis of the scales for the appropriate track and then the guideline charging rates for that particular court or group of courts. Different rates are applied depending upon the professional status and experience of the lawyer conducting the case, with partners attracting the highest and trainees and paralegals the lowest, and employed solicitors and legal executives between. The rates are only for guidance, but judges tend to adhere to them fairly rigidly.

There may be a shortfall between the rates applied and what is charged by the solicitors to their clients and between the overall amounts allowed on assessment and the total bills to the clients. Indeed, shortfalls are almost inevitable. That is why so many landlords, managing agents and management companies are now conducting their own litigation — especially if they do not possess the bargaining power to come to some arrangement with their solicitors.

When parties act for themselves for all or part of any court proceedings they may be entitled to recover "litigant in person" costs. Since the arrival of the CPR, "litigants in person" have included companies (although charities must be legally represented).

Where costs are awarded to a litigant in person, there are various rules and limitations affecting what can be allowed:

- apart from disbursements, a litigant in person may not be awarded more than two-thirds of the figure which would have been allowed if the party had been legally represented
- the litigant may recover what he has laid out for legal services relating to the proceedings
- for individual items of timed work, the litigant can recover either what he can show he has lost for that time or, failing that, an allowance for the time spent at (currently) £9.25 per hour
- a litigant who is allowed to claim the costs of attending court may not recover a witness allowance for the same hearing.

Managing agents will no doubt consider whether it is either appropriate or economic for them to offer a representation service to their clients as part of an overall "added value" project. If they do, their fees are a matter between them and their client. There is little chance of their fees being recoverable from their client's opponent.

Costs may be assessed in one of two ways: summary assessment or detailed assessment. A detailed assessment is more akin to the old method of "taxation": a formal and detailed bill of costs is prepared and served on the paying party. If objections are raised, the matter goes before a "costs judge" (usually a district judge) who carries out the detailed assessment at a hearing set aside for that purpose. If no objections are made within the time allowed, the successful party is entitled to draw up a default costs certificate which, when issued by the court, is enforceable as any other money judgment.

Particularly in lower level and more straightforward cases however, judges are encouraged to carry out a summary assessment, which is particularly well suited to interim applications and fast track trials. It may also be of use in small claims if costs are to be awarded. A summary assessment normally takes place immediately at the end of the hearing and it is inevitably somewhat rough and ready. The broad brush approach can be brutal (perhaps especially for the sensitive advocate) but it has the distinct advantages of (a) avoiding the additional costs of the detailed assessment procedure and (b) achieving certainty at the earliest possible opportunity.

If an order for costs is to be sought at a hearing and it is the type of hearing where summary assessment is possible (which is practically any hearing in a debt recovery context) the parties need to be ready. In particular, this means that a costs statement or schedule should be served on the other side at least 24 hours before the hearing. County court form N260 is available for this purpose. The form is not mandatory but production of a costs statement in some form is compulsory in fast track trials and interim applications. Arguably, it is not compulsory in small claims but it is good practice if costs orders are to be sought. In any event, the penalty for failing to serve and produce a costs statement is that the court will almost certainly disallow some or all of that party's costs.

The costs statement is intended for legal representatives, but litigants in person should use their best endeavours to modify it for their needs.

Contractual costs

If a landlord or management company has incurred costs which are not put forward to the court for assessment, the only other alternatives for recovery are under the defaulting leaseholder's covenants in the lease or through the service charge from all contributing leaseholders.

Essentially, the points to be made here are the same as in the forfeiture context (see Chapter 2 and elsewhere). Costs can only be charged to the extent that they are reasonable and if the express wording of the lease permits it. If there is any doubt, it will be resolved in the leaseholder's favour. The onus is on the landlord or management company to prove that the relevant clause gives authority.

The chances are high that any dispute on this issue will end up in the LVT even though the costs were incurred in the county court, because the LVT will be the forum for the majority of cases on service charges and administration charges.

Appeals and setting aside judgments

Appeals are rare in debt recovery cases so it is not intended to go into them in detail here. As with the LVT, permission to appeal is needed either from the judge whose decision is to be appealed or, failing that, from the judge of the appellate court. Appeals from a district judge go to a circuit judge for example. The time-limit for lodging an appeal in the county court is 14 days, although this may be extended on application.

Permission to appeal is only likely to be given if the appellant has a real prospect of success in arguing that the decision was either wrong in law or was arrived at unjustly because of a serious procedural or other irregularity.

Not to be confused with an appeal, a judgment may also be disturbed by an application to set it aside. These usually arise with default judgments or decisions made in a party's absence. Unless the judgment had been entered wrongly, the court will usually require the applicant to demonstrate a real prospect of success in the proceedings, and he may be ordered to comply with conditions (such as making a payment into court). The court is also entitled to take into account whether the application was made promptly.

8 Enforcement of Repairing Covenants

In most of this volume so far the focus has been on breaches of covenant for non-payment of monies, whether they be rent or service charges. That is no accident; the vast majority of enforcement cases are to do with non-payment. Other covenants are broken from time to time however, and it would be a very fortunate property manager who never has to deal with one. Because they are less frequent, such enforcement cases are perhaps less well understood. They tend to be more expensive, more time-consuming and generally more convoluted — they also tend to require more professional input, whether from lawyers, surveyors or others.

This chapter will consider the actions available on a breach of a repairing covenant. From the property manager's point of view, enforcement will generally be in the context of a leaseholder's covenant, but enforcement of landlords' covenants will be examined briefly in Chapter 12.

What constitutes a breach of a repairing covenant?

The nature of a covenant was described in Chapter 1: in particular, the wording of a covenant must be clear and unambiguous to be enforceable, and the onus to establish the correct interpretation is on the person seeking to enforce it. This task is no easier with repairing covenants than with any other type; indeed, huge volumes of decided legal cases turn on interpreting the respective repairing obligations under a lease. Even now, many issues remain unclear. There is no space here to repeat the arguments, but some points to watch out for are:

- is the lease clear that the area in disrepair falls under the leaseholder's responsibility (which is not necessarily the same thing as being within the demise)?
- is the area or item out of repair within the lease's definition of what the leaseholder is to repair?
- does the item come within the judicially accepted definition of "structure"?
- does the lease require any preliminary steps before enforcement — such as a period of notice?
- even if it is clearly a breach, does the disrepair affect the landlord or any third parties detrimentally (see under Leasehold Property (Repairs) Act 1938, below)?

If in doubt, take legal advice on interpretation or a surveyor's advice on the proper description of the item out of repair.

The lease may require that the leaseholder is served with a schedule of dilapidations or notice of wants of repair. Whether it is a requirement under the lease or not, it is generally good practice to serve one and give the leaseholder a reasonable time to comply. The costs of preparing such a schedule or notice will usually be recoverable from the leaseholder under the terms of the lease, but obviously this should be checked.

Determining the breach

Although forfeiture may have become less popular in the context of recovering debts, it is still seen as the best method for achieving results on other breaches of covenant. As we shall see below the alternatives are quite limited.

Forfeiture for breach of repairing covenants is subject to the same restrictions as other breaches (see Chapter 2), particularly as a result of sections 168 and 169 of the Commonhold and Leasehold Reform Act 2002 ("the 2002 Act"), which came into force on 28 February 2005. As a reminder of the burden of those sections:

- a landlord may not serve a notice under section 146 of the Law of Property Act 1925 (a "section 146 notice" — the preliminary to forfeiture) unless either of the next two points is satisfied.

- the leaseholder has admitted the breach of covenant concerned or
- the breach has been determined to have occurred by a court, the LVT or arbitration
- even then, a further 14 days must have passed after the determination and any time for appeals and the breach must still be unremedied
- the only arbitration to qualify will be under a post-dispute arbitration agreement: an arbitration clause in a lease will be void for this purpose.

Assuming that the leaseholder has not admitted the breach (if he has, skip to "section 146 notice", but hopefully an admission would lead to the situation being remedied anyway), it will be necessary to have the breach determined before taking further action. This will be expensive and will take a significant period of time. In an emergency, landlords should consider seeking an injunction or (if the lease empowers them to do so) carry out the repairs themselves and sue for reimbursement afterwards. If the case is being taken up on behalf of other leaseholders, landlords may wish to seek their indemnity as to the costs before commencing proceedings. (The latter issues are all discussed separately later in this book.)

The next question to decide is whether to apply to the court or the LVT to make the determination (it is likely to be extremely rare that the parties agree to enter into a post-dispute arbitration agreement).

Section 168(4) of the 2002 Act says:

> A landlord ... may make an application to a leasehold valuation tribunal for a determination that a breach ... has occurred.

It does not say that the application can only go to the LVT so, presumably, it is still permissible to apply to the county court. There are advantages and disadvantages either way, and those considerations may even vary geographically. For example, in some areas the local court may be quicker generally than the LVT region, while in others the truth will be the opposite. Perhaps the biggest argument for going to court is the greater chance of being awarded costs; conversely, the risk of going to court is that the court may take the view that the LVT should have been selected in the first place and transfer the case there, occasioning further delays and expense and possibly penalising the landlord in costs.

It is important to remember however that the landlord is only entitled to recover his costs if they are provided for under the lease or if they are awarded as costs of the proceedings. The court may do the latter; it is rare for the LVT to do so. Meanwhile, whether there is a contractual entitlement will depend upon the precise wording of the lease: if the clause relates specifically to a section 146 notice and no notice can be served until after the determination, it is doubtful that the landlord can recoup the costs of the determination process.

It is going to be difficult to develop a general rule for choice of forum. Much may depend upon the degree to which the landlord relies on lawyers to take such cases forward. In service charge cases, there is a high risk of transfer to the LVT, but this is probably much less so with other types of breach where historically the court has an expertise which the LVT has yet to build up. It could well be that the courts will be more ready to transfer repairs cases to an LVT whose members include a surveyor than they will with more technical issues such as unauthorised assignments or sub-lettings or even nuisance cases.

As repairs cases will be relatively rare, the choice of forum is something for which few property managers need to have a routine, but the decision on each individual case should be considered carefully. (An application to the LVT will be substantially similar to the service charge scenario outlined in Chapter 6 and Appendix 3. A case study for an application for a determination by the court will appear in Chapter 9.)

Section 146 Notice

If the breach is admitted or once it has been determined and requisite time has expired (and assuming it remains unremedied) the landlord is free to serve a section 146 notice. The notice can only be served by the landlord: if the manager does not possess that capacity (for instance as a management company which does not own the freehold), the full co-operation of the landlord will be necessary, which may involve the prior giving of an indemnity as to the landlord's costs.

The following scenario gives an example, based upon the case study in Appendix 1.

Scenario 4: Flat 12A Tumbledown Mansions

The leaseholders of Flat 12A are a young couple, Lou and Sue Hoodoo. They are prone to accidents. Recently, Lou was training with weights when he tripped on a loose floorboard, causing him to lose his balance.

The bar-bell went through the window (breaking the glass and buckling the metal frame) and landed at one end of the balcony attached to the flat. The weight was such that the balcony became unfixed and it is now dangling precariously over the balcony of Flat 8 below. Mr Witt of Flat 8 is a director of Tumbledown Management Ltd ("TML"), the management company for the block.

The insurers have declined to meet the claim.

TML instructed its managing agents, Reddy & Willing, to pursue the matter. There is no argument that both the window frame and the balcony are demised to Flat 12A and Lou and Sue admit that they covenant to put and keep them in repair. Unfortunately, they are unable to raise the funds to pay for the repairs.

Reddy & Willing's staff surveyor, Rick Fellows, prepares and serves a schedule of dilapidations in accordance with the lease, but Lou and Sue are no further forward. Indeed, as a result of the injuries he sustained at the time of the accident, Lou is now off work on statutory sick pay. To make matters worse, Sue has lost her job, so their financial position is worse than ever.

TML agree with the freeholder, Reston Laurels plc, that Reston Laurels will instruct TML's solicitors (Swift & Sharp) to serve a section 146 notice upon Lou and Sue (who have already admitted the breach) on condition that TML will be responsible for the costs of the matter, in so far as there is any shortfall. The section 146 notice prepared by Swift & Sharp appears at Appendix 5.

The notice must be served in accordance with section 196 of the Law of Property Act 1925 by recorded delivery or personally. Personal service can include insertion through the letter-box of the premises. Copies should also be sent to any mortgage lenders or subtenants with an interest in the property as they will be affected by any subsequent forfeiture and they are entitled to apply for relief from forfeiture (see Chapter 2).

Leasehold Property (Repairs) Act 1938

As mentioned earlier in this chapter, forfeiture for breach of a repairing covenant is subject to the same restrictions as other breaches; moreover, it has an additional restriction all to itself. The Leasehold Property (Repairs) Act 1938 requires that any section 146 notice must inform the leaseholders of their right to serve a counter-notice under the Act. If the leaseholders serve a counter-notice within 28 days the landlord can take no further steps on the notice without the leave of the court.

The court will only grant permission if the landlord can prove on the balance of probabilities that at least one of the grounds in section 1(5) is satisfied. The grounds are as follows:

- the breach must be remedied immediately to prevent substantial diminution in the value of the freehold reversion or the value has already been diminished substantially
- the breach must be remedied immediately to comply with statute, the requirement of any authority or byelaw or a court order
- the repair is necessary in the interests of another occupier (where the leaseholder is not in occupation)
- the cost of an immediate repair is relatively small in comparison with the probable expense caused by postponement
- there are special circumstances rendering it just and equitable for the court to give leave.

In the event of Lou and Sue serving a counter-notice, it is likely that Reston Laurels would succeed on the last ground if on no others, relying on the health and safety aspect (as far as the balcony is concerned at least).

The forfeiture proceedings

If no counter-notice is served or if leave is granted, the forfeiture is effected by the issue of possession proceedings in the county court local to the property. The proceedings are prepared fundamentally in the same way as in a forfeiture for non-payment of rent (see Appendix 2) with obvious consequential changes to reflect the basis of the claim. The particulars of claim should also include a claim for damages for breach of covenant in an unspecified amount. It is likeliest that the court will order the defendants to carry out the necessary repairs and make good so that the only compensation will be in costs, so damages will probably not be awarded; nonetheless, they should be claimed.

Along with the lease and office copies of the entries at HM Land Registry, copies of the schedule of dilapidations and the section 146 notice should be annexed to the particulars of claim. If the court or LVT had determined the existence of the breach then a copy of their decision should also be referred to and annexed.

Relief from forfeiture

The general principles of relief from forfeiture are discussed in Chapter 2. In this scenario, relief from forfeiture is likely to be granted on terms that:

- the breaches of covenant are remedied by the necessary repair works being done within a time scale agreed between the parties or fixed by the court
- the defendants pay the claimant's costs to be the subject of detailed assessment if not agreed within, say, 28 days of assessment or agreement
- upon compliance the defendants to be relieved from forfeiture
- if the defendants fail to comply with the previous points, the claimant to be entitled to possession.

If the party being granted relief were actually the defendants' mortgage lender, things would have to be dealt with slightly differently depending upon whether the defendants were to remain in possession. If the defendants are staying, the simplest thing would be for the lender to be joined as an additional defendant to the forfeiture proceedings and then to undertake to deal with the first two points of the order. They would then seek to recoup their outlay by adding it to the defendants' mortgage.

Alternatively, it sometimes happens that the lender is content to wait until the claimant landlord has recovered possession as against its borrowers, and then to step in and obtain relief on the basis of similar terms, but with the additional point that an order is made vesting the leasehold title in the lender. In practice in such circumstances, the lender becomes the leaseholder itself (often on the basis of a new lease — but without paying a premium) for such time as it takes to carry out the necessary repair works and compensate the landlord in costs. The lender (as "mortgagee in possession") then sells the property, recouping its losses and redeeming its loan from the sale proceeds. Any shortfall will become a personal debt due from the original defendants to the lender.

It is rare for a lender to intervene in this way, but it does happen, especially if the leaseholders were in substantial or flagrant breach of their lease and their mortgage deed simultaneously.

Costs of the forfeiture proceedings

Because a forfeiture for disrepair will probably have been dealt with as a multi track case, the scales of costs allowed can be more generous. In an extreme case, the Claimant landlord may even have been awarded costs on an indemnity basis. In practice however, the difference between the indemnity and standard bases is not vast, and the chances are that costs assessed by the court will create a shortfall. Also, the costs of this case will have been heavy. The complexity of the issues and possibly the law involved (if there were arguments over interpretation of the lease for example) would ensure high fees on the part of the lawyers, and there will have been significant surveyor's fees as well. Accordingly, a relatively small percentage shortfall in costs recovered will mean that the landlord (and thus the management company) being considerably out of pocket. Whether the shortfall could be put to the service charge will depend upon the wording of the lease, but it would not be popular in any event.

On top of that, the court may have refused to allow anything at all for the costs of any LVT proceedings or for work done before the service of the section 146 notice.

In all the circumstances, the landlord and management company are likely to be better off if they can negotiate with the defendants or (more probably) their mortgage lenders for payment of their costs by way of an overall compensatory amount for the original breach.

As a passing comment, it is perhaps a little odd that a landlord (especially a residents' management company answering that description) is more likely to be able to achieve by negotiation rather than court order a conclusion which fits more closely the wording of section 146(3):

> A lessor shall be entitled to recover ... all reasonable costs and expenses ... in reference to any breach giving rise to a right of re-entry or forfeiture ... from which the lessee is relieved, under the provisions of this Act.

Unless and until the wind changes direction again, it is prudent for landlords and managers to keep an eye on ways of minimising costs, and lawyers should be pushing hard to achieve the fullest possible indemnities when the terms for relief from forfeiture are being considered.

Alternative remedies for disrepair

There are other ways of responding to a leaseholder's breach of a repairing covenant. The court has power to order a party to do or not to do something (an injunction or an order for the specific performance of a contractual obligation). Meanwhile, landlords have certain common law powers in certain circumstances to carry out repairs to the leaseholder's premises and recover their expenditure afterwards.

Injunctions and specific performance

A landlord or a management company can apply to the court for an injunction and/or a decree of specific performance. Such an application can be made on its own or as part of possession proceedings; alternatively the application could proceed while the LVT is determining the existence of the breach. Most logically perhaps, the application could be made together with an application to the court to determine the breach.

The meaning of an injunction is well known and understood, perhaps because it is widely used in a variety of civil disputes such as those between husband and wife. The same cannot be said of the remedy known as "specific performance". This remedy is only available to the parties to a contract and it is intended for use if one party is in breach of his obligations to the other and where no other legal remedy (such as damages) is adequate to protect the injured party's position. The term "specific performance" is self-explanatory: the court may order one party to the contract to perform a specific contractual obligation. In that sense, it is an order positively to do something whereas an injunction is usually an order not to do something.

Applications for remedies such as injunctions or specific performance are either made as part and parcel of other proceedings (such as possession claims or claims for declarations — which would be the path for an application for the court to determine the existence of a breach) or as the principal issue under part 8 of the CPR. A part 8 claim for a declaration will be examined in Chapter 9. There are some particular considerations to take into account when applying for injunctions or decrees of specific performance.

- Such remedies cannot be obtained by default. A hearing will be necessary and the applicant must have a strong case.

- Although the injunction may be the main point of the proceedings, they should generally be combined with other issues such as a claim for damages for breach of covenant, even though there may be no serious intention to pursue the damages claim (at the outset at least).
- Breach of an injunction is a contempt of court which may be punishable by imprisonment in extreme cases; accordingly, the court will only grant an order if entirely satisfied that it is appropriate and proportionate.
- Evidence should be of a high standard, fully and frankly disclosing all material facts. Affidavits or affirmations will generally be required, rather than merely witness statements verified by statements of truth. The conduct of both sides is relevant to the exercise of the court's discretion.
- Injunctions are usually intended to be of short duration, protecting the parties' relative positions or preserving the status quo while other issues between the parties are being resolved (such as a forfeiture claim).
- The party seeking the injunction will usually be required to give an undertaking to compensate the other party against any losses incurred in the event of an eventual judgment that the injunction was misconceived.
- Not all levels of judge in the county court have jurisdiction to grant every type of injunction. It is wise to check with the court when issuing the application to save wasted time and cost later. (Applications made within proceedings for possession of land or which have been allocated to the small claims or fast track are within the district judge's jurisdiction.)
- An injunction can be obtained before the issue of the substantive proceedings in urgent cases, but they will usually be subject to an undertaking to issue the claim as soon as possible.
- Generally, the applicant will be expected to have a draft injunction available for the court's approval.
- Injunctions and similar orders are only effective once they have been validly served.
- The court may allow witnesses' addresses to be withheld from defendants, or even their names and evidence in very extreme cases.

Injunction applications are hard work and the need to get things right is even more imperative than usual. Applications for specific performance are, if anything, more difficult simply because of their rarity. It is very unusual for parties to attempt such applications without legal representation.

These points are recognised by the courts however, and successful applicants tend to find that the court will be more generous in awarding costs than in more straightforward litigation. Although the costs are inevitably higher, there is a better prospect of maximising recovery. As in all matters however, the court will look to be proportionate and will be less favourable therefore if there is any doubt over the absolute necessity of the application or whether there were any ulterior motives on the part of the applicant.

The rights of other leaseholders

It is not only the landlord or management company who can seek an injunction or even specific performance. In our scenario above, the leaseholders of flat 8 could seek an injunction or damages in the event of any loss being suffered or (with strong evidence) to prevent imminent damage being caused to their flat. More to the point though would be if the leaseholders of flat 8 could apply for an order requiring a positive act on the part of the leaseholders of flat 12A.

In exceptional circumstances, the court will grant a mandatory injunction requiring that something be done rather than not done. The leaseholders of flat 8 would have a stronger case though for specific performance if they can demonstrate a contractual obligation on the part of the leaseholders of flat 12A.

Many leases contain covenants by leaseholders which are made expressly with the landlord and the other leaseholders of the development. Conventionally, such covenants are those the performance of which impinges most directly on other residents, and typically they will include the leaseholders' repairing covenant. If this is the case, then it is open to other leaseholders (such as Mr and Mrs Witt of flat 8) to apply to the court in their own right for a decree of specific performance. Indeed, their application is likely to receive a more sympathetic hearing than that of the landlord.

In such circumstances, it is often advisable for the landlord and the other affected leaseholders to be joint applicants (although of course the leaseholders could not be joint claimants to a possession action).

The landlord's powers of entry to execute works

Although there is an implied covenant by the leaseholder to permit entry to the landlord if the landlord is responsible to repair the premises, it is important to distinguish that from the more normal situation where the leaseholder is responsible to repair the demised premises. Here, an express covenant is required.

None the less, it is not at all uncommon for a lease to contain a provision enabling the landlord to enter a demised flat to inspect the state of repair and, in certain circumstances, to carry out repairs himself.

The provision may be found in a number of places, usually within the leaseholder's repairing covenant itself or sometimes within the schedule of the exceptions or reservations from the demise. The wording and extent of this provision vary considerably, but common elements include:

- the landlord must give reasonable notice of an inspection except in an emergency
- a requirement that the leaseholder carries out repairs notified to him by the landlord within, say, three months of notification, failing which the landlord is empowered to enter to execute the works
- the resulting costs and expenses will then be recoverable either as debt or damages. (If suing for damages, the landlord must first give notice under the Leasehold Property (Repairs) Act 1938.)

Any attempts by the landlord or manager to gain access to the leaseholder's premises without lawful authority under the lease or otherwise could lead to civil proceedings for trespass and breach of the covenant for quiet enjoyment and possibly a criminal prosecution for harassment or wrongful eviction under the Protection from Eviction Act 1977. It cannot be stressed enough that no landlord, management company or property manager should embark on this course without the reassurance of independent legal advice that the lease actually means what they think it means.

9 Enforcement for Other Breaches

What other covenants may need to be enforced?

Previous chapters have dealt with enforcement of the leaseholder's covenants to pay rent and service charge and to carry out repairs. The answer to the question of how many other covenants may have to be enforced by property managers will depend upon what other covenants exist in a particular lease. Several of the most frequently breached covenants are described below, but there could be many others. Generally, the options and procedures for enforcing covenants in long residential leases are essentially the same unless they fall into the categories set out in previous chapters (rent, service charges and repairs) which have special rules peculiar to them.

Administration charges

Administration charges tend to be considered in the context of how and when they might be recovered as part and parcel of an enforcement action for another breach. It is quite likely however that they may need to be collected under a covenant in their own right, either because they have arisen in an entirely different manner or because the breach of covenant which brought them on has been remedied, leaving only the charges outstanding.

The term "Administration Charges" was defined by paragraph 1 of schedule 11 to the Commonhold and Leasehold Reform Act 2002 as follows:

1 (1) In this Part of this Schedule "administration charge" means an amount payable by a tenant of a dwelling as part of or in addition to the rent which is payable, directly or indirectly —

(a) for or in connection with the grant of approvals under his lease, or applications for such approvals,
(b) for or in connection with the provision of information or documents by or on behalf of the landlord or a person who is party to his lease otherwise than as landlord or tenant,
(c) in respect of a failure by the tenant to make a payment by the due date to the landlord or a person who is party to his lease otherwise than as landlord or tenant, or
(d) in connection with a breach (or alleged breach) of a covenant or condition in his lease.

By restricting the definition to charges payable "as part of or in addition to the rent" the schedule would seem to apply only to payments contractually due under the lease. Consequently, some charges would appear to escape, most obviously the fees raised by managing agents for responding to solicitors' conveyancing enquiries. These are not payable under the lease, but rather under a separate one-off contract with leaseholders' solicitors (notwithstanding that it will be the leaseholder who pays in due course). Indeed, if the enquiries are made by an intending purchaser who has yet to acquire the lease, there is no landlord and tenant relationship between the parties at all.

It is not yet clear whether subparagraphs (c) and (d) might be construed to include such things as legal costs in enforcement actions. There will usually be contractual provision in the lease to pay the costs (including solicitors' and surveyors' fees) for such matters, but legal costs are also subject to assessment by the Courts (and now by the LVT to a certain extent) and it may be found to be more appropriate to deal with costs separately. None the less, the wording seems to suggest that legal costs should be encompassed within the definition.

It can be assumed safely that interest payable under a lease in the event of late payment of rent or service charges would be included within subparagraph (c).

Parliament has expressly reserved to the Secretary of State the power to amend (and therefore extend) the definition of administration charges (paragraph 1(4) of schedule 11).

Administration charges are payable "only to the extent that the amount of the charge is reasonable" (paragraph 2, schedule 11 of the 2002 Act). Determinations of reasonableness are made by an LVT.

Demands for an administration charge must be accompanied by a summary of leaseholders' rights and obligations relating to such charges (paragraph 4, schedule 11). Although a prescribed form for these summaries has yet to be introduced, the statutory requirement to serve them came into force on 30 September 2003, leaving property managers to invent their own forms for the time being. (Some of the professional and trade bodies have developed precedents. Alternatively, it has been suggested that the model draft at Annex D of *A Consultation Paper on Accounting for Leaseholders Monies and summaries of tenants rights and obligations* should be used. Copies can be downloaded from *www.odpm.gov.uk*.)

If no accompanying summary has been served, a leaseholder may withhold payment, and none of the penalties in the lease for late or non-payment will apply (paragraph 4(3) and (4), schedule 11). Consequently, no enforcement is possible unless and until a summary has been served.

Subject to the last point, and to the fundamental consideration that it must be validly recoverable under the lease, there are two principal routes for enforcing payment of an administration charge:

- debt recovery through the county court (essentially the same process as that outlined in Chapter 7)
- forfeiture for breach of covenant, subject to prior determination by the court or more probably the LVT (very much along the same lines as the service charge case described in Chapter 6).

Unauthorised alterations

It is common for leaseholders to covenant not to make structural alterations to their flats either at all or without the landlord's prior consent. Enforcement of this covenant can be required urgently if an unauthorised alteration threatens to undermine the structural integrity of the building and is quite likely that an injunction will be sought. Superficially, an enforcement of this covenant appears similar to that for the repairing covenant (see Chapter 8). However there is no role here for the Leasehold Property (Repairs) Act 1938.

Having said that, a leaseholder carrying out an alteration may well have infringed his repairing covenant as well. The general rule is that enforcement action should hit as many targets as possible so, as long as no prejudice is caused by additional delay, it would probably be best to proceed simultaneously on the alterations and repairing breaches, with regard as necessary to the requirements of the Leasehold Property (Repairs) Act 1938 set out in Chapter 8.

Assignment and subletting

It is unusual for residential long leases to carry any prohibitions against the leaseholder's rights to assign the lease or sublet his flat except in the last few years of the term. More frequently, although probably still in a minority of cases, there may be conditions upon such events, particularly in retirement schemes or where the leaseholder holds a share in a residents' management company. For example, the leaseholder may covenant not to assign the lease without entering into a licence to assign. It is quite common for there to be a prohibition against assigning or subletting just part of the demised premises.

These points tend to be picked up by the solicitors or licensed conveyancers acting for the intending purchaser of the flat, but occasionally a breach occurs. If it does, it will usually be in the interests of the leaseholder to have the matter resolved as quickly as possible (especially if a mortgage lender is involved), so correspondence or a section 146 notice at most should suffice. If further action is required however, the threat of forfeiture is likely to be the most effective. The procedures are largely as those described more fully in the scenario below.

Registration

Unfortunately, the covenant to register assignments and sublettings with the landlord or management company (which is found in almost every lease) does not meet with compliance nearly often enough. The same can be said of the regularity with which pre-contract conveyancing enquiries are raised of the manager, but that does not give rise to a breach of covenant.

It is not unheard of for notices to be registered a year or more after the assignment or subletting took place. This can cause considerable administrative difficulties and inconvenience to managers: for example, consultation procedures may go awry and cash flow can be seriously dented. On the face of it however, failure to serve notice of assignment seems a relatively trivial matter — at least, that is usually the standpoint of the solicitor or conveyancer who failed to serve the notice. Perhaps as a result of that, arguments over late registration tend to be among the most heated.

If it comes to it, the forfeiture approach is still the most suitable for enforcing this covenant, but patience and tolerance are recommended in practice.

Immoral or illegal user

Proceedings against residential leaseholders regarding the way in which they use their demised premises are very rare, although occasionally they happen. Examples of a breach of the covenant not to use the premises for any immoral or illegal purposes would include using the flat as a brothel or for the distribution of illegal drugs. Such activities could also be caught by covenants against causing nuisance or annoyance or those barring carrying on a business.

The difficulty with pursuing these cases is that of proof and finding willing witnesses. Record of a criminal conviction makes it very much easier, but in practice the police are not always keen to prosecute (no doubt they see advantages in knowing where to keep an eye on things from the point of view of containment). In the event that a case were proved, clearly the breach of covenant in this context would fit into the category known as "irremediable breaches" in that, once the breach has occurred, it could not be remedied. Nevertheless, this is not necessarily a bar to relief from forfeiture. So long as the breach stops and can be reasonably expected not to recur and the landlord can be compensated the leaseholder (or those deriving title under him such as mortgage lenders or subtenants) is quite likely to be granted relief by the court; it all depends on the merits and developments of the particular case.

(A case study looking at a case of this type in more detail can be found in Chapter 10.)

Insurance with the landlord's nominated insurer

In most residential leases, particularly for flats, it is the landlord who covenants to maintain the buildings insurance and charge the costs on to the leaseholders, either separately from or as part of the service charge accounting. Either way, payment of insurance contributions is governed by the same legislation as for service charges. Some leases however provide for the leaseholder to effect the insurance, but through the landlord's nominated or approved insurer. This applies most frequently to leasehold houses and maisonettes.

Parliament perceived the practice of taking forfeiture proceedings for breach of this covenant as an abuse, especially if leaseholders had arranged perfectly adequate insurance cover but not through the landlord's insurer (whose premiums might be considerably higher — particularly when the landlord's commission was taken into account). The Commonhold and Leasehold Reform Act 2002 did not outlaw this approach, but it has made it much more difficult in two ways:

- it is now possible for leaseholders of houses to avoid the burden of such covenants, provided they follow the "notice of cover" requirements of section 164 of the 2002 Act
- the landlord now has to go through the procedures to have the breach determined before he can serve a section 146 notice, thus giving the leaseholder more time to resolve the problem and (crucially) requiring the landlord to run the risk of incurring considerable costs which he is unlikely to recover.

Observing regulations

Most modern leases now differentiate between "covenants" and "regulations" or "stipulations". Forfeiture is available for breach of covenant, but not for the less fundamental matter of the breach of a regulation. Most leases get around this distinction however by containing a covenant by the leaseholder to observe the regulations and stipulations (which will usually be listed in a separate schedule). Accordingly, enforcement action can be taken for breach of a regulation by way of forfeiture — although it rarely is unless there are additional breaches of covenant which justify the proceedings.

It is typical for leases containing regulations to have a provision reserving the power to the landlord or management company to amend or introduce new regulations. Some are tempted to use this power to bring in quite draconian measures without having to go through the process of obtaining variations to the leases. The temptation can be quite overwhelming: for example, the directors of a management company of a prestigious residential development might be understandably concerned about the impact on their amenities and the values of the flats on the estate if a lot of subtenants started to appear. As a protective measure, the directors could put through a regulation requiring leaseholders to obtain prior consent to subletting or imposing conditions on the terms of tenancies.

It is highly improbable that a court would uphold an attempt to forfeit a lease where a leaseholder had failed to observe such a fundamental regulation — especially when he was not aware of it when he entered into the lease. Of course, the behaviour of subtenants may lead to forfeiture on entirely separate grounds, such as breaches of the leaseholder's covenant not to permit nuisance or annoyance to be caused to other residents.

Nuisance

Apart from arrears and repairs cases, probably the most frequent complaints of breach of leaseholders' covenants revolve around nuisance and annoyance issues. Often, but not always, the activities giving rise to complaints are those of subtenants. Nuisance and annoyance can come in a wide variety of forms. Unfortunately, the variety of forms of words in leases is almost as wide, and great care needs to be taken to ensure that the covenant actually covers the act or acts in question.

Nuisance complaints inevitably tend to lead to the most heated disputes and the most entrenched positions on all sides. It is important for the property manager and the professional advisers to stand back and take a detached look at the overall situation and the particular complaint in the light of the specific covenant concerned. For example, does the complaint relate to behaviour which emanates from outside the flat when the covenant expressly relates to activities within the demised premises? (This is not an unusual scenario.) If so, there is little that the landlord or management company can do about it.

If there is an arguable case that a breach of covenant has occurred, then forfeiture is usually the most suitable approach, but property managers may be best advised to assess the alternatives which, depending upon the precise circumstances.

- If the activities causing the nuisance are actionable in themselves, then the complainant can take his own action, and this option deserves to be considered at least.
- Is the complaint borne of mutual disaffection, so that the landlord will end up caught in the middle of a "neighbour dispute? If so, is there an arbitration clause in the lease for such disputes? If there is, it should be invoked. (Arbitration clauses for disputes as

between leaseholders are not rendered void by the Commonhold and Leasehold Reform Act 2002 — see Chapter 12.)
- Is the landlord obliged to take action by a covenant in the lease? If so, is he entitled to require an indemnity for his costs before doing so? It is wise to rely on such a provision nowadays.
- Is there adequate available evidence to back up any proceedings? In particular, are there enough independent witnesses who will be willing to attend court?
- Is the matter so pressing that an injunction should be applied for? If so, the application should be coupled with some other claim, possibly forfeiture, but that would be premature before a section 146 notice has been served. Timing needs to be considered carefully.
- If the problems are caused by subtenants, the obvious step of contacting the leaseholder and requiring him to sort out the problem should not be overlooked. It may be that he should be pushed to enforce his tenancy agreement by taking the preliminary steps towards forfeiture; indeed that may help him to persuade the court to order possession. On the other hand, the landlord needs to be aware that he may not recover the costs of precipitate action, so timing again becomes an important consideration.
- Is the situation such that civil proceedings are actually the less appropriate strategy? Should the problem be referred to other agencies such as the police or the local authority, who may be able to use powers in the context of such things as noise pollution or anti-social behaviour?
- Are there human rights or discrimination issues to be addressed?
- Generally, should the matter be referred to the landlord's or management company's lawyers for advice?

Many of these points will be examined in greater depth in the case study in Chapter 10.

Determining the breach

The purpose of what follows is to illustrate the standard procedures involved in an enforcement case without getting bogged down in the more convoluted questions wrapped up in a nuisance case. The situation is again drawn from the case study at Appendix 1.

Scenario 5: Flat 5 Tumbledown Mansions and garage

Some of the leases for Tumbledown Mansions (including that for flat 5) include a garage within the definition of the demised premises. These leases also contain a covenant by the leaseholder:

> Not at any time during the term to assign underlet or otherwise part with possession of part only of the demised premises as opposed to the whole.

The leaseholder of flat 5 is Mr A Wohl who has sublet the flat to Mr and Mrs T Dance. Mr and Mrs Dance did not require use of the garage however and Mr Wohl (through his agents Play A Let) let the garage separately to Mr Christopher (known as Kit) Carr.

The garage demised with flat 5 is not the only one at Tumbledown Mansions to have been sublet separately, but the directors of the management company, TML, feel they cannot ignore it in this case as Mr Carr has begun to run an *ad hoc* car repair service from the garage, spilling out on to the forecourt and access ways and causing congestion.

When challenged, Mr Carr says he will try not to get in anyone's way, but he is only doing the odd favour for a friend. The directors are not satisfied and they contact Mr Wohl's agents. The agents say they will pass TML's comments on to Mr Wohl, but make no other comment. After three weeks and follow-up calls to Play A Let, there is no sign of any change or of any attempts to terminate the tenancy to Mr Carr.

TML discuss their options with their managing agents (Reddy & Willing) and consult their solicitors (Swift & Sharp). They also obtain authority from the freeholder, Reston Laurels plc, to embark upon forfeiture proceedings in their name if necessary subject to a complete indemnity as to the freeholder's costs.

Choosing the forum for determining the breach

Since nothing has been heard from Mr Wohl, clearly the breach has not been admitted by him, neither has anyone purported to admit it on his behalf. TML then must choose how to have the breach determined, by the court or the LVT.

They decide to apply to the court for three reasons:

- the court has well established rules for dealing with parties who are resident abroad

- there is a much greater chance of recovering costs
- if the case is defended, any defence is likely to be based on legal argument rather than a dispute over the facts (namely interpretation of the lease and whether TML has waived its rights to enforce this covenant by deciding not to take action against other leaseholders on the same facts).

The application to the court

The court is being asked to determine a question concerning the applicant's legal rights. This is known as a "declaratory judgment" or more commonly a "declaration". Unlike injunctions, the court may make a declaration whether or not any other remedy is claimed (rule 40.20 of the CPR). A claim for a declaration is made under part 8 of the CPR. A sample Claim Form N208 and supporting particulars of claim for this case can be found at Appendix 6. The forms can be modified for use in service charges cases and declarations relating to other breaches of covenant (except non-payment of rent, for which this process is not relevant).

The same points apply to service of the claim upon Mr Wohl as those discussed in the scenario contained in Chapter 5.

Because part 8 claims are intended for cases where there is no factual dispute of any substance, there is no corresponding form for a defence or admission sent out to the defendant, only a form for acknowledging service (N210). This form gives the defendant a limited number of options (apart from ignoring it of course). It should be returned within 14 days of service, and the defendant can:

- confirm that he does not intend to contest the claim
- confirm that he does intend to contest the claim
- indicate an intention to challenge the court's jurisdiction
- object to the claimant's use of the part 8 procedure.

If the defendant intends to respond in any of the contentious ways referred to, he should file his supporting evidence at the same time as the acknowledgement of service or within 14 days afterwards. The claimant may serve evidence in reply 14 days after the defendant's evidence has been served upon him. The time limits can be extended by agreement between the parties (but only by an additional 14 days for the defendant's response) or by the court.

Challenges to the court's jurisdiction generally relate to geographical considerations, but there is no reason why a defendant should not argue that a leasehold matter rests properly with the LVT. Since the court's and the LVT's jurisdictions are parallel however, this argument is unlikely to succeed (which is not to say that the court will not transfer the case to the LVT for convenience or other reasons in due course).

An objection to the use of the part 8 procedure is likely to be on the grounds that there is in fact a substantial dispute of fact. Even if the defendant does not apply, the court will move to prevent an abuse of the system if it becomes obvious. Exceptionally the claim might be struck out, but more usually it will be redesignated as a part 7 claim and then be allocated to a track in the usual way, with the claimant possibly being penalised in costs. If a residential leasehold case is indeed revealed to involve such a substantial factual argument, that is one of the possible reasons for the court to transfer the proceedings to the LVT.

If the defendant fails to respond in any way to the claim except to attend the hearing, the rules state that he may not participate without the court's permission. In practice, permission is usually given so long as the defendant has a good reason for his earlier silence or he shows due contrition.

Part 8 claims are automatically allocated to the multi-track, so there is no need for an allocation stage. The court will give directions (including fixing the hearing date) either when the claim is issued or when the acknowledgement of service has been filed. Local practices vary.

The hearing

Most part 8 claims can be dealt with at a short hearing largely on the basis of the filed evidence, especially if the defendant raises no serious dispute. Generally in such circumstances no oral evidence will be required, but it is prudent to have the witnesses in attendance in case the need for clarification arises or cross-examination is permitted.

In our case, neither Mr Wohl nor anyone on his behalf files any response of any sort, and the claim is heard in his absence by District Judge Scales. The claimant is represented by Mr Sharp of Swift & Sharp (the claimant's solicitors). Ms Reddy of the managing agents is also present. Mr Sharp introduces the matter and refers the judge to the claim form and supporting documents; he also confirms that he has checked with Ms Reddy that the problem was continuing on the morning of the hearing and that Ms Reddy can give that in evidence if required. The judge says that she is satisfied and orders in the terms set out in the claim form ("order as prayed" is lawyers' shorthand phrase for such an outcome).

The rest of the hearing turns on the question of the claimant's costs. Mr Sharp produces a copy of a statement of costs which he filed with the court and served upon the defendant 48 hours previously. (See Chapter 7 under "Assessment of costs".) The costs statement sets out his time on the case overall at his normal charging rate as a partner in the firm and that of his assistant (a trainee legal executive), together with the court and Land Registry fees incurred on the claimant's behalf. The firm's hourly charging rates are in accordance with the court's guideline rates for costs assessments.

The statement also includes an element of time charges for Ms Reddy as the claimant's agent in attending court and generally assembling the evidence and liaising with the solicitors. Anticipating some resistance from the judge on this point, Mr Sharp argues that the claimant's costs claim is based primarily on its contractual entitlement in the lease. He quotes the leaseholder's covenant to pay

> all costs charges and expenses (including solicitors' and surveyors' fees) incurred by the Lessor ... arising as a result of the breach of any of the Lessee's covenants...

and submits to the judge that the clause is drawn widely enough to encompass the fees of the freeholder's managing agents.

The judge listens attentively to Mr Sharp, but says that while she understands his point she does not see this as a case for using the court's discretion so as to allow costs on any more favourable terms than in normal proceedings. She has the power to award the claimant's witness her expenses and a sum up to £50 to compensate for provable loss of earnings. No proof of either was offered and, in any event, the judge imagines that such activities are part and parcel of the managing agent's function for which they agree a fee with their clients.

As to the solicitors' fees, the judge remarks that there looked like some possible duplication of work between Mr Sharp and his assistant, and in any event she thought that too much time had been taken in preparing the claim form and particulars of claim. The judge then ordered that the claimant's costs claim should be allowed after deduction of the managing agent's fees and 20% of the solicitors' charges.

Following the determination

The order is processed by the court and served on the parties. In accordance with sections 168 and 169 of the Commonhold and Leasehold Reform Act 2002, Mr Sharp works out that Reston Laurels plc will have to wait for 28 days before serving a section 146 notice. This calculation is based on the time-limit for appeals in the county court (14 days) plus a further 14 days to be allowed under section 168(3) of the 2002 Act. He makes a diary note and advises Reston Laurels and TML (through Reddy & Willing) accordingly.

Meanwhile, in accordance with the terms of the agreement between Reston Laurels and TML, Mr Sharp renders his firm's interim account to Reddy & Willing who pay the bill and allocate it 80% to the account for flat 5 and 20% cent to the service charge for the block.

After 28 days have expired, Mr Sharp checks the up to date position with Reddy & Willing (nothing has changed) and obtains confirmation from Reston Laurels that he should proceed to prepare and serve a section 146 notice (see Appendix 6).

Following service of the section 146 notice and, if necessary, the implementation of the forfeiture of the lease by the commencement of possession proceedings, it is to be assumed that Mr Wohl or his representatives would eventually contact Reston Laurels and TML to negotiate a settlement and obtain relief from forfeiture.

The terms for relief from forfeiture in this case are likely to include:

- an undertaking to terminate the tenancy for the garage (of which Mr Carr is in breach anyway) as a matter of urgency (the terms should include a timescale and make relief conditional on its successful completion)
- an undertaking not to breach this covenant again
- clearing any other breaches which may have occurred by this point, such as any service charge arrears
- payment of the costs of the part 8 proceedings as ordered
- payment of any subsequent costs in relation to the section 146 notice and any forfeiture proceedings (it may be possible here to include a negotiated allowance for the managing agent's fees incurred since the declaration since there may not be any need to obtain a court order — other than perhaps a consent order — for agreed terms for relief from forfeiture).

A note on undertakings

It is not infrequent for terms for relief from forfeiture to include undertakings to perform covenants in future or carry out certain acts to remedy outstanding breaches. Arguably, the undertakings may be unnecessary, since the covenants are enforceable in any event. Nevertheless it is useful to obtain them, not least because they concentrate the leaseholder's mind on his responsibilities.

Whenever possible and appropriate it is preferable, for the party benefiting from the undertaking, that the undertaking should be given to the court. The undertaking then becomes enforceable in its own right, in the event of a breach, by an application for committal to prison as a contempt of court. No doubt because of the serious consequences for breach, the court has stringent requirements for the formalities of an undertaking, and they will be determined by the particular circumstances and type of case. It is sensible to take legal advice when framing undertakings, although court officials can also assist.

Injunctions

Injunctions are dealt with in both specific and general terms elsewhere in this volume. In the particular scenario set out above, it is unlikely that either Reston Laurels plc or Tumbledown Management Ltd would have been granted an injunction simply on the covenant regarding parting with possession. They may have been successful if some other covenant had also been breached for which an injunction was more appropriate. It is more likely however that another leaseholder whose right of way was being interfered with would be able to obtain an injunction against the subtenant rather than Mr Wohl.

Costs

Costs are discussed in a number of chapters; for example Chapter 7 looks at assessment of costs by the court and Chapter 13 contains a general examination of the subject. In the scenario above, the freeholder and management company have been relatively successful, but even so they incurred a shortfall of 20% of the solicitors' bill which has been put to the service charge.

It may well be that that would be an end of the matter. If the solicitors' charges for the part 8 claim were, say, £1,000 (which is by no means intended to be read as a "going rate") then the shortfall would be £200. Spread across 20 flats, each leaseholder would contribute £10. It is conceivable that the charge could be challenged by any leaseholder on a number of grounds, such as:

- the lease does not authorise such payments from the service charge
- the figure is unreasonable — if it had not been the court would have allowed it
- (on Mr Wohl's part) how is the shortfall recoverable from him, even in part, when the court had already disallowed it?

Obviously the risk of a challenge would be commensurately greater if the shortfall was greater.

There is no easy answer to this dilemma; it will depend on a range of circumstances, most particularly the terms of the lease. From the point of view of managers of leasehold property, even this relatively successful scenario points to the need to take all possible steps to keep costs to a minimum.

10 Neighbour Disputes

The purpose of this chapter is to look at the situation whereby a landlord or management company is drawn into taking an active part in disputes between neighbouring leaseholders, or complaints by one or more leaseholders against another resident. Sometimes, that active part is to act as or appoint an arbitrator as between leaseholders under an arbitration clause in the parties' leases. The mechanics of arbitrations are discussed in Chapter 12.

Thus it would be more accurate to say that this chapter examines when landlords have to enforce covenants against leaseholders at the behest of other leaseholders in the context of the sort of complaints typically categorised as "neighbour disputes":

- noise
- nuisance and annoyance
- "anti-social behaviour"
- trespassing upon a leaseholder's peaceful enjoyment of his flat and the associated rights
- acting to the detriment of the amenities of the estate and the value of the flats.

Many of the legal and strategic issues raised by such cases are described in Chapter 9, particularly under the headings of "Nuisance" and "Immoral or Illegal User". The main points are set out below.

- Neighbour disputes tend to be very heated. Detached, independent advice should be obtained before proceeding.

- Is the problem caused by subtenants? If so, the simple step of an amicable approach to the leaseholder should not be overlooked.
- These cases can be very expensive and, looked at callously, there is likely to be no benefit from them for landlords (of whatever description). Landlords and management companies are advised to insist on indemnities for their costs from complainants if the lease allows it.
- The facts complained of must be clearly in contravention of the wording of the covenant relied upon.
- Should application be made for an injunction?
- Should the matter be referred to others such as the police or local authorities?
- Cases can easily fail because of insufficient evidence or unwilling witnesses.
- An "irremediable breach" does not guarantee that the leaseholder will be barred from relief from forfeiture.

The scenario which follows is an effort to illustrate the complexities of such a case and to point to some of the available options for managers. For background, please refer to the case study at Appendix 1.

Scenario 6: Reston Laurels plc v String

Reddy & Willing, as managing agents for TML, have received a number of complaints concerning Ms G String, the leaseholder of flat 15, Tumbledown Mansions. The most regular and vociferous complainant is Mr Lars Straw of flat 10 (the flat below flat 15). Mr Straw complains of Ms String making noise and causing a nuisance. Lars works shifts at the gym and his sleep is disrupted by Ms String and friends in the early hours. He cites loud music, lots of banging and his light fittings shaking. Lars has complained to Ms String but she relies on her right to use her flat as she sees fit and the spirit of free enterprise, and in any event she denies that she is causing a nuisance. She suggests that Lars is ultra-sensitive and recommends an anger management course.

Lars is at the end of his tether and demands action. He asserts further that this potential dispute may jeopardise his sale (he is anxious to move because of the ongoing noise nuisance) and he threatens to instruct his solicitors to claim damages from TML if his sale falls through.

Lars is not the only complainant however. Reddy & Willing and directors of TML (when they meet to discuss the matter) find that complaints have been made orally or in writing from:

- Mrs Woolley of flat 2 (close to the main door of the block) who is concerned at the number of night-time visitors and the noise they make. She is frightened that they may try to gain access to her flat for burglarious or other nefarious activities.
- Miss Keen of flat 7. Miss Keen is a young woman who says she has been harassed on the stairs by male visitors to the block on a number of occasions.
- Mrs Witt of flat 8 says she has had similar experiences and she is pushing her husband (a director of TML) to do something about it.
- Mr and Mrs Nutt of flat 9 make general complaints of noise.
- Fallon Branch of flat 14 is a director of TML and she confirms that she has been "solicited" by men calling at the wrong door and that, on occasion, it is very noisy in flat 15 (which is next door). She also says however that the problem is intermittent and that, generally, she gets on very well with her neighbour Ms String.
- Miss Barrass of flat 19 has told Mr Martinet that she is extremely upset at the thought of what is going on. She has written him a number of lengthy but confidential letters on the subject.
- Mr Rockett of flat 20 makes regular complaints regarding noise from flat 15, which is immediately below his flat. He says he has called the police on a number of occasions, but they seem to do nothing about it.

The chairman of TML, Mr Martinet, also expresses personal concern. His flat is next to the main front door of the block and his window gives a full view of the car park. He states that what is going on inside is nothing compared to nocturnal behaviour in the car park, although he has been unable to identify the individuals involved as the lighting in the car park has been interfered with.

Given the volume of the complaints, the directors of TML decide that action must be taken, but the nature of the case is such that they feel the need to tread carefully. They ask Ms Reddy to instruct TML's solicitors to consider the alternatives and advise TML upon them. Ms Reddy puts together a detailed bundle of documentation to go to Mr Sharp of Swift & Sharp (TML's solicitors) including copies of the written complaints, details of the oral complaints she knows about, and a copy of Ms String's lease.

Two weeks later, Mr Sharp replies with a lengthy letter advising a two way approach.

1. Representatives of TML should contact the local police and request firmly that they investigate what appears to be a brothel being run in the block. TML should offer full co-operation with providing evidence and facilities for the police to keep close scrutiny. They should also check the block and particularly the car park for any evidence of drugs use, as this may help the police decide to give the matter a higher priority. Mr Sharp suggests as an aside that individual residents could make direct approaches to the police to seek an Anti-Social Behaviour Order (ASBO). He doubts that it would be very effective in this case but it may be worth trying. The police interest may have the desired effect anyway, but if not a record of conviction would make a forfeiture action much easier.
2. TML should also approach the freeholder, Reston Laurels plc, and ask them to initiate forfeiture action for breaches of covenant. Mr Sharp's letter identifies a lengthy list of covenants which may have been breached, but also points out that some will be more difficult to prove than others. His letter also sounds a note of caution to the directors, emphasising how messy and expensive these cases can be and especially the difficulty of coming up with adequate evidence.

The directors of TML instruct Reddy & Willing to do what they can to put Mr Sharp's advice into effect. Reston Laurels respond in their usual way in such circumstances by authorising TML to proceed to instruct Swift & Sharp to take the necessary steps towards forfeiture so long as TML gives Reston Laurels a full indemnity as to any costs they may incur or that they may be ordered to pay. They also suggest employing an enquiry agent to collect evidence for the case to make up for understandable reluctance on the part of residents to get involved. Apparently they have used this approach before in a similar case. Finally, they say that, in their experience, little active co-operation is to be expected from the police.

Meanwhile, nothing is heard officially from the police for a considerable time, although Ms Reddy receives a mysterious "off the record" call from a Detective Constable Wayne Hay. DC Hay tells her that there are ongoing police inquiries affecting Tumbledown Mansions. The police are aware of Ms String's activities but they are not the main focus of their attentions. All he would say is that the property is of more interest to the drugs squad than the vice squad and that they would prefer not to prejudice other matters by showing an obvious police presence at the property for a little while. Ms Reddy reports back to the directors as discreetly as she can.

Strategy

In all the circumstances, the TML directors decide that they have little alternative than to press on, particularly as some of the complainants are becoming very restive. They are conscious however that they have not yet put the allegations to Ms String in any formal way and they anticipate that the court will expect there to have been at least a letter before action. Furthermore there has clearly been no admission of the breaches of covenant and an application to the court or LVT to determine the breach will be necessary. They do not expect Ms String to admit the breaches but they know she must be given the opportunity.

All this will take time, and would involve giving Ms String forewarning of at least some of their evidence. The directors are not confident that all the complainants will be willing to attend court in any event. Before sections 168 and 169 of the Commonhold and Leasehold Reform Act 2002 came into force in February 2005, it would have been permissible to proceed straight to a section 146 notice, perhaps served personally by an enquiry agent while he was gathering supporting evidence.

A meeting is convened with the directors of TML, Ms Reddy from the managing agents, Mr Sharp from TML's solicitors and Mr B Eddie Vere (who, jointly with his brother Percy, is the principal shareholder in Reston Laurels plc). The meeting results in a plan of campaign:

1. Each of the complaining leaseholders would be requested to make and sign a witness statement, confirm their willingness to attend court, and make a payment of £500 on account of the costs of the case and as a sign of commitment (in accordance with the clause in the lease whereby the lessor can insist on security for costs in consideration for acting upon a leaseholder's complaint).
2. Subject to receiving at least three such responses, Swift & Sharp would instruct their regular enquiry agent to prepare an independent report confirming (or not) the leaseholders' allegations and adding any other relevant information.
3. Swift & Sharp would then put all the papers together and instruct counsel to draft an application to the court for a determination of the breach or breaches coupled with an application for an injunction preventing Ms String from continuing her breaches or

permitting others to act in such a way as to cause nuisance or annoyance to leaseholders or other residents.

4. Counsel will also be asked to advise upon whether Ms String should be given a letter before action and, if so, when. Counsel is to be invited to draft an appropriate letter.
5. If necessary after the breaches had been determined, a section 146 notice was to be served and possession proceedings issued as soon as reasonably permissible thereafter.
6. Swift & Sharp should keep the police up-to-date at every point and co-operate and liaise with them as necessary. Swift & Sharp are given authority to soft pedal on the forfeiture case if it becomes apparent that the police are taking their own action against Ms String.

It is agreed at the meeting that the application for determination of the breaches will be made to the court for similar reasons to the case in Chapter 9 (recoverability of costs, the possible need for interpretation of the covenants, the tighter rules for service of documents and summoning witnesses and so forth) but also because of the need for the injunction to be dealt with simultaneously within the same proceedings. In this type of case, timing is everything.

The proceedings

Essentially, the claim for a declaration will follow the same structure and path as the part 8 claim outlined in Chapter 9 and Appendix 6. Obviously the particulars of claim and the supporting witness statements will require considerably more evidence than on the more technical single issue in Chapter 9. The same evidence will also have to support the claim for an injunction.

(There is little point in attempting a precedent for an injunction application, since such claims are largely determined by the facts of the particular case and barristers and solicitors vary considerably in their preferred drafting methods. The essential rules and principles for applications are described briefly in Chapter 8.)

Evidence

Swift & Sharp instruct their usual process server and enquiry agent, Samuel Shovel, to gather what evidence he can concerning the activities of Ms String. His report confirms that Ms String has a few regular male callers, usually in the evenings, who stay for varying lengths of time but never all night. Her visitors seem reluctant to be recognised. He has also pieced together some information from the internet and "Contact" magazines that indicates that Ms String also goes by the name of "Wellbeloved" and that she advertises personal services of an expensive nature involving equipment traditionally used by escapologists and lion-tamers.

When Sam calls to see Mr Sharp to deliver his report, he tells him confidentially that certain other things came to light during the course of his inquiries which might be of interest to Mr Sharp. First, he thought he recognised some of Ms String's callers, including some local dignitaries. Second, the nocturnal visitors to the parking area seemed to be visiting not Ms String but another resident of Tumbledown Mansions altogether; furthermore, the purpose of their visits appeared to be purchasing items in small plastic bags rather than anything more intimate. Third, he had noticed that he was not the only one keeping a trained eye on the estate, but the others were watching the car park rather than the building itself. Finally, Sam had heard, entirely confidentially and off the record and not to be repeated to anyone in any circumstances whatsoever, that the interest of the police focused on a resident of Tumbledown Mansions who worked in the local pharmacy. Obviously, Sam did not want any of this information appearing in his evidence, and Mr Sharp was not to pass it on under any circumstances.

After some consideration, Mr Sharp decides to concentrate simply on the main issue of Ms String and that, notwithstanding Sam's intelligence on the car park visitors, there was enough evidence from all available sources to convince a judge that Ms String was at least causing nuisance and annoyance to other residents. He also believes that there is a strong chance that the judge will accept that there is enough to establish, on the balance of probabilities (which is the test in a civil action), that Ms String is in breach of other covenants including illegal or immoral user and running a business from the demised premises.

In this somewhat civilised scenario, there are sufficient willing witnesses to give live evidence in support of the claim. Often, unfortunately, the situation on the ground in reality is much more unpleasant and willing witnesses are few and far between. In a 2004 case, some comfort was forthcoming for landlords and management companies when the available evidence was less than perfect.

In *Solon South West Housing Association Ltd* v *James* [2004] EWCA Civ 1847, the use of hearsay evidence was allowed (and this was supported on appeal) in a possession case where the reason live evidence from hearsay witnesses was not adduced was their fear of reprisals. The judge took the hearsay nature into account when weighing the evidence, and the hearsay evidence fitted the pattern of the live evidence (which had taken priority).

Although this is a reassuring case, there still had to be some live evidence. It is in such circumstances that the use of "professional" witnesses such as enquiry agents or (in appropriate cases) local authority noise pollution officers can be so useful.

The outcome

The proceedings are taken as planned, with Reston Laurels plc and TML as joint claimants on this occasion. An interim injunction is obtained at an early stage and it is served upon Ms String by Sam. She accepts service in a cheery mood, and tells Sam that she is pleasantly surprised to have got away with things for so long.

Ms String does not contest the proceedings and, shortly afterwards, she quietly moves out. She has a mortgage over the flat and her lender negotiates a settlement with Reston Laurels and TML through Swift & Sharp, whereby the lender takes possession of the flat under its mortgage deed and discharges the costs of the case in full. The flat is sold in due course to a respectable semi-retired gentleman, Mr M D Sade.

Anti-social behaviour orders

In the case of local authorities and social landlords, the new ASBOs and injunctions (ASBIs) will afford considerable assistance for dealing with so-called "nuisance tenants". These provisions are likely to be of far less use from a practical point of view as regards long leaseholds and the private sector, unless the nuisance emanates from tenants of such landlords.

In so far as there is a remedy available under this legislation, it will be of benefit to individual residents rather than landlords or management companies of long leasehold blocks. ASBOs have come about through a complicated web of legislation: one definition for example is from the Crime and Disorder Act 1998 as amended by the Police Reform Act 2002 and the Anti-Social Behaviour Act 2003:

> An application for an order under this section may be made by a relevant authority if it appears to the authority that the following conditions are fulfilled with respect to any person aged 10 or over, namely:
>
> (a) that the person has acted, since the commencement date, in an anti-social manner, that is to say, in a manner that caused or was likely to cause harassment, alarm or distress to one or more persons not of the same household as himself; and
>
> (b) that such an order is necessary to protect relevant persons ... from further anti-social acts by him;

Sufferers should be referred to the social landlord concerned, the local authority or the police. Only these organisations can apply to the courts for ASBOs or ASBIs. More information can be obtained from a number of sources, including local authorities and the Office of the Deputy Prime Minister (*www.housing.odpm.gsi.gov.uk* or *www.odpm.gov.uk*).

11 Enforcement Outside Leases

This book concentrates on enforcement of covenants in leases, but the manager of leasehold property may sometimes have to take action against parties, other than leaseholders, but who none the less have an interest in the managed property. This could include the freeholder, but enforcement against landlords is considered in Chapter 12. More often the issue will crop up in relation to owners and occupiers of residential units not held under leases. As with leaseholders, the predominant issue will be contributions into a service charge fund, but other questions will arise less frequently, generally in relation to the amenities and common parts of the estate.

Freehold units

There has been an increasing trend in new residential developments for the types of construction to be mixed. A typical example would be an estate consisting of one or more blocks of flats and a separate group of individual houses all using common amenities such as estate roads and visitors' parking areas. The flats would be leasehold but the house-owners would have freehold titles. The transfer deed or other form of title document for each house would contain covenants on the part of the first purchaser and provisions binding successive owners of the house, just like a lease, but crucially there would be no forfeiture clause.

Put simply, a lease is for a fixed term of years but at its termination the property reverts to the landlord (hence the "freehold reversion"), and a forfeiture clause allows for early termination on breach of covenant. A freehold title continues into perpetuity so there is no reversion to the landlord, and thus no right of early termination.

So, if the landlord or management company of the estate cannot threaten a freehold house-owner with forfeiture for breach of covenant, what other weapons are there?

Essentially, the answer is that all the same enforcement measures are available against freeholders as against leaseholders with the single exception of forfeiture — and as a result the LVT has no role to play (except in very limited circumstances: see under "Estate management schemes" below). A brief list of options would be:

- debt proceedings for unpaid contributions
- damages for breach of covenant
- injunctions preventing breaches of covenant
- specific performance.

Procedurally, such measures are effectively identical to those set out in Chapter 7 (for debt recovery) and Chapters 8 and 9.

Generally, service charge contributions payable by freehold house owners and their other obligations to an estate owner are known as "rent-charges", which is a misnomer since rent does not come into the equation with a freehold unit at all. Sometimes, covenants will refer instead to "estate service charges" or "estate rentcharges" — indeed the latter is really the more proper term since "rentcharges" strictly were abolished by the Rentcharges Act 1977.

Costs

It is not uncommon for freehold unit-holders to covenant to pay the estate owner's or manager's costs in the event of enforcement of their other covenants or rentcharges. The prospects of actually recovering such contractual costs are about the same as with similar provisions under leases, except of course that the additional layers of case law and statute surrounding costs in a forfeiture context will be irrelevant.

In short, it is always worth claiming contractual costs, but the chances of recovering them are slight unless they are incurred in proceedings of a type whereby the costs of the action are normally awarded.

Estate management schemes

Where leasehold properties have been enfranchised under the Leasehold Reform Act 1967 or the Leasehold Reform, Housing and Urban Development Act 1993, the former service charge structure may have been translated into an estate management scheme as part of the enfranchisement process.

Section 159 of the Commonhold and Leasehold Reform Act 2002 provides that estate service charges under an estate management scheme are only payable to the extent that they are reasonable. The LVT is the appropriate forum for determining any issues of reasonableness. The Tribunal also has the power to vary any formula in the scheme documentation to achieve a reasonable arrangement.

Commonholds

By the time of writing it was not thought that any commonhold developments had been completed, so any discussion of a commonhold association's powers of enforcement against a unit-holder is necessarily academic.

The commonhold association's rights to enforce against a unit holder are set out in the prescribed regulations for commonhold associations laid down by The Commonhold Regulations 2004 ("the regulations") (SI 2004 No 1829). The regulations contain at schedule 3 the (Commonhold Community Statement — the rules of the development "CCS"). Paragraphs 4.11.11 to 4.11.16 contain the key passages. They are sandwiched between the rules for a unit holder to enforce against the association and those by which one unit holder may enforce against another.

If the complaint against a unit holder is that he has defaulted in a payment or that his breach of a duty under the CCS creates an emergency, the commonhold association may issue legal proceedings, refer the matter to the Housing Ombudsman, or use the dispute resolution procedure in paragraphs 4.11.12 to 4.11.16 of the regulations. Other issues should go through the dispute resolution procedure.

The procedure consists of several stages.

1. The commonhold association must first consider resolving the matter by negotiation, arbitration, mediation, conciliation or another form of alternative dispute resolution other than legal

proceedings, or taking no action if that is in the best interests of the unit-holders as a whole.

2. If it wishes to proceed, the commonhold association must serve a default notice upon the unit-holder. The notice is in a prescribed form. The defaulter may respond, also by use of a prescribed form.
3. On receipt of a reply or 21 days after the default notice, the commonhold association must again consider the same factors as at point 1.
4. If the matter is still unresolved, the commonhold association may either refer the matter to the Housing Ombudsman (if the commonhold association is a member of the relevant scheme) or issue legal proceedings, but only if the commonhold association considers such action serves the best interests of the development.

After the above procedures have been exhausted, any legal proceedings brought will be essentially the same as those that could be taken against a freehold house-owner in an otherwise leasehold estate.

There is one notable additional power in favour of commonhold associations however: if the commonhold unit is let to a tenant, the commonhold association may require the tenant to divert rent payments to the commonhold association to discharge a debt.

12 Arbitration and Alternative Dispute Resolution

Historically, and especially before the advent of the LVT in the service charge context, the only alternative to court proceedings was arbitration when it came to disputes between landlords and leaseholders. Even then, the possibility of arbitration only arose realistically when the lease contained a clause providing for disputes to be referred to arbitration.

The historical balance has been upset considerably over the last few years, both by the intervention of the LVT and by the official encouragement of other forms of alternative dispute resolution, "ADR" — principally mediation. Meanwhile, arbitration clauses as between landlord and leaseholder (but not otherwise) have been undermined by the Commonhold and Leasehold Reform Act 2002.

Arbitration

In the event of a dispute between the parties to an arbitration agreement (such as an arbitration clause in a lease) it is generally mandatory to invoke the agreement by referring the matter to an arbitrator. It has been held in other areas of contract law that failure to pay or otherwise meet a contractual obligation amounts to a dispute for the purpose of establishing whether an arbitration clause should be implemented.

The objective of an agreement to refer a dispute to arbitration is to achieve a resolution of the dispute by an impartial forum (usually a single arbitrator) with relative informality and as little inconvenience and expense as possible. An arbitrator's decision generally could be appealed only on grounds that the arbitrator had erred in law or manifestly misconducted the proceedings. The decision would be binding on the parties but on no-one else. It would be very unusual for a decision to be published. The parties tend to agree the arbitrator's terms of reference, but it is usually for the arbitrator to lay down the way in which the procedure is conducted and the award of costs (including determining who pays the arbitrator's fees).

Arbitration clauses in leases

Most arbitration clauses in residential leases provide for the parties to agree the appointment of the arbitrator or, in default, to apply to the President of the Law Society or Royal Institution of Chartered Surveyors to nominate a suitable person. Some clauses appoint the landlord's surveyor as the arbitrator, which could make things look just a little one-sided.

The other obvious failing of arbitration clauses is that they were entered into by the original leaseholder. Successors in title often are unaware of the existence of such a clause and therefore can hardly be said to have entered into it willingly.

Nevertheless, until the implementation of the relevant sections of the Commonhold and Leasehold Reform Act 2002, the presence of arbitration clauses in leases ousted the jurisdiction of the LVT in disputes covered by the clause in question (always assuming that the parties and their advisers had noticed them). That situation changed fundamentally in February 2005 when sections 168 to 170 of the 2002 Act came into force (see below under "Post-dispute arbitration agreements and the 2002 Act").

Arbitration between leaseholders

Some leases contain provisions to resolve disputes between leaseholders (or sometimes between leaseholders and/or other residents) by way of an arbitration process. In such cases the arbitrator is often the landlord or management company, or their surveyor or agent. The arbitrator generally determines who pays whose costs and how his fees are apportioned as part of his decision.

Arbitration clauses of this type survive the Commonhold and Leasehold Reform Act 2002. They can provide a useful alternative to litigation for leaseholders. They can be something of a double-edged sword for landlords and managers however: while providing for a relatively tidy mechanism for killing off disputes between neighbours, they also tend to draw the manager into the dispute. Landlords and managers who would rather avoid being caught by such a dilemma might like to consider pleading a conflict of interest and declining the appointment.

Post-dispute arbitration agreements and the 2002 Act

Reference has been made above to the changes made to arbitration agreements between landlords and leaseholders by the Commonhold and Leasehold Reform Act 2002. The Act deals with arbitration at a number of different points, but the cumulative effect is as follows:

- for practical purposes, any dispute between landlord and leaseholder can be determined by the court, the LVT or arbitration (the Act refers to an "arbitral tribunal")
- the only valid form of arbitration in this context is under a post-dispute arbitration agreement
- any other form of arbitration agreement (including an arbitration clause in a lease) will be void for such disputes.

A post-dispute arbitration agreement

> means an arbitration agreement made after the breach has occurred (or is alleged to have occurred)

(section 168(5) of the 2002 Act) and it must be in accordance with part 1 of the Arbitration Act 1996.

Arbitration Act 1996

Section 1 of the 1996 Act lays down the following principles of arbitration:

> (a) the object of arbitration is to obtain the fair resolution of disputes by an impartial tribunal without unnecessary delay or expense;
> (b) the parties should be free to agree how their disputes are resolved, subject only to such safeguards as are necessary in the public interest;
> (c) in matters governed by this Part of the Act the court should not intervene except as provided by this Part.

The Act also contains certain requirements of a valid arbitration agreement.

- It must be in writing.
- Any legal proceedings commenced by a party to an arbitration agreement should generally be stayed.
- The parties are free to agree the number of arbitrators, but an agreement to appoint an even number will imply that an additional member should be appointed to the panel to chair it. If there is no agreement, there shall be a sole arbitrator.
- If the parties cannot agree the appointment, any party may apply to the court to make the appointment.
- The parties shall be jointly and severally liable for the arbitrator's fees and expenses (subject to any award of costs — see below).
- Procedural and evidential matters are generally up to the arbitrator to decide.
- The arbitrator has much the same powers as the court when it comes to what can be awarded (although enforcement will be through the court).

Appeals

The circumstances in which an arbitrator's decision can be overturned are still limited under the Arbitration Act. The relevant points are contained in sections 68 to 70.

A decision may be challenged in the court under section 68 for a serious irregularity if:

- the arbitrator has failed to comply with his duties under the Act
- the arbitrator has exceeded his powers
- he has failed to follow the procedures agreed by the parties
- he has failed to deal with all the issues before him
- the award is uncertain or ambiguous
- the award was obtained by fraud
- the arbitrator has admitted any irregularity.

Meanwhile, under section 69, an award may be appealed generally only on a point of law (as opposed to a question of fact), and then only with the agreement of the parties or the permission of the court. Section 69 goes on to set out the factors the court must take into account when considering whether to give permission to appeal. Sections 68 and 69 both give details of the variety of decisions a court can make on a successful appeal, including setting aside the award, varying it, or sending it back to the arbitrator for further consideration.

Section 70 of the 1996 Act contains a number of supplementary provisions concerning appeals, including most importantly that any appeal must be brought within 28 days of the award.

Any parties to an arbitration who are considering challenging or appealing the award are advised to go through the full terms of sections 68 to 70 in detail.

Costs of the arbitration process

Sections 59 to 65 of the Arbitration Act 1996 set out the arbitrator's powers to award costs. In summary, an arbitrator may make an award as to costs, but only to the extent that costs are reasonable and if they are properly recoverable under the terms of the arbitration agreement itself (which, for our purposes, is not the clause in the lease of course but the post-dispute arbitration agreement).

A strategic decision will have to be made when agreeing the terms of any post-dispute arbitration agreement when it comes to whether to insert express provision regulating awards for costs. The successful party (or the party anticipating that happy position) may take the view that it would be preferable not to tie the arbitrator's hands. This is the sort of policy issue which should be discussed with the landlord's or management company's legal advisers in the event that the arbitration approach is one which merits exploration.

Mediation and conciliation

Mediation or conciliation may not have a direct role in enforcement *per se*, but that is not to minimise their importance. If a dispute can be mediated or conciliated to a successful, amicable conclusion then either there will be no need for enforcement or (if the leaseholder fails to fulfil an agreement) there may be an admission on record which could avoid the need to determine the existence of a breach of covenant. Assuming the mediation or conciliation process can be completed swiftly and economically, then landlords and management companies will be better off than they would have been had litigation been necessary, at least in theory.

Mediation is perceived as particularly useful for retirement and sheltered housing, where litigation can be considered as distasteful. Age Concern England offers mediation through AIMS (Advice, Information and Mediation Service). A number of other mediation services are available across the wider market, but the professional services of an accredited mediator can be expensive, possibly disproportionately so for small points or low value disputes. Because the mediator has no decision-making role, there is no basis for any awards of costs. Consequently, even before deciding whether mediation is appropriate for a particular dispute, research needs to be carried out on availability and cost.

Mediation and conciliation differ from litigation (whether through the court, the LVT or arbitration) in a variety of ways. Sometimes the differences will be advantageous, other times they will be disadvantageous. Each dispute or category of dispute needs to be looked at individually to determine whether such forms of ADR will be of use. For example:

- litigation leads to a binding decision (unless the case is withdrawn): mediation only achieves this if all the parties agree
- most litigation (except arbitrations) is in public and decisions are on the record: neither applies to mediation
- although relatively few court cases get to a hearing, there is always the risk of aggressive, adversarial hearings with cross-examination of witnesses: parties to ADR need never face each other or any legal representatives
- litigation requires the parties to set out their case at the outset; later changes of course can be penalised and can be seen as weakening the party's position: mediation carries no such burden.
- attempts at ADR are encouraged generally by the courts: in some situations, courts can order the parties to attempt mediation.

Most practitioners lean against ADR for most enforcement cases for three central reasons: irrecoverable costs; additional delay in achieving a resolution; no assurance of a binding decision. Also, the cynical practitioner would say that if an amicable solution were likely, there would have been no enforcement necessary. That may be a blinkered view, but it is accurate to say that very few service charge cases are settled by mediation. It may however have more of a role in cases involving other types of breach of covenant, such as repairs or nuisance.

Further, ADR as between leaseholders is something which many property managers would encourage. It may be sensible to suggest ADR as a first step when one leaseholder complains about another and requests the landlord or management company to take action. The suggestion could be all the more persuasive when coupled with information regarding the landlord's right to security for costs from the complainant before taking any action.

Enforcement against landlords

This book is written principally in the context of enforcement by landlords (however defined) against leaseholders. That is partly because most of those active in managing leasehold property (whether as professionals in the field or as directors of residents' management companies or right to manage companies) will think of enforcement primarily in the context of leaseholders' covenants. Second, all the recent legislative activity has focused on the landlord's enforcement powers. (It is worth remembering at this point that "landlords" in a service charge context include "any person who has a right to enforce payment of a service charge" (section 30, Landlord and Tenant Act 1985) — therefore residents' management companies, Right to manage companies and tight to enfranchise companies are encompassed within this definition.)

Landlords and management companies enter into covenants as well, and they also need to be enforced from time to time. Indeed, on occasion management companies need to take enforcement measures against freeholders and vice versa. For example, leases to which management companies are parties ideally should contain some mechanism for ensuring continuity of performance of management obligations if the management company becomes insolvent or otherwise fails. Usually, this mechanism will require the freeholder to step in and take over the management role with such remedies as may be of practical use against the management company — or possibly its directors.

Covenants typically given by landlords and management companies are described in Chapter 1. Those which are most commonly encountered in enforcement actions would be:

- (by the landlord) to insure, repair, maintain and redecorate the structure and common parts; to give quiet enjoyment; to enforce against other leaseholders
- (by the management company) to insure, repair, maintain and

redecorate the structure and common parts; to collect and account for service charges.

Actions to enforce covenants against landlords and management companies are brought by way of claims to the court for damages and/or injunctions and/or decrees of specific performance. The way these remedies work is discussed at length elsewhere in this volume, albeit in the reverse context of enforcement against leaseholders. None the less, the procedures are the same as are the more strategic considerations surrounding such matters as costs, and issues such as the need to ensure that the express covenant actually requires the landlord to do what the court is being asked to order him to do.

Covenant for quiet enjoyment

There is one of the landlord's covenants which is worth highlighting, because it is unique to the landlord and because it is not generally well understood, and that is the covenant for "quiet enjoyment". This covenant precludes the landlord from interfering with the leaseholder's possession of his property and the lawful use and enjoyment of its amenities. The most extreme breach of the covenant would be wrongful eviction or harassment (both of which are criminal offences under the Protection from Eviction Act 1977).

How far the covenant extends was considered in two cases which got as far as the House of Lords in 1998/99. In the cases of *Southwark LBC* v *Mills and Baxter* v *Camden LBC*, the tenants occupied flats which had been constructed with no sound insulation and they suffered from noise emanating from flats let to other tenants of the same landlord. The landlords were held not to be liable either for breach of quiet enjoyment or for authorising a nuisance.

The way that covenants and their meanings can run into one another is exemplified by the case of *Chartered Trust plc* v *Davies*. This case involved elements of nuisance, and the covenant for quiet enjoyment, and the landlord's obligation to take up the cudgels for one tenant against another, but it actually centred on the landlord's covenant not to derogate from his grant. This is another typical covenant by which the landlord promises not to do or allow anything which will detract from what he has granted in the lease or tenancy. The facts of the case involved the tenancy of shop premises so they are not particularly relevant *per se*, but the point at issue was that the landlord was held to be in breach of the covenant not to derogate from his grant because that covenant imposed a duty to prevent a nuisance from other (and newer) tenants from rendering the claimant's premises unfit for the purpose for which they had been let.

The role of the LVT

There is an alternative way of enforcing the covenants and statutory obligations of landlords and managers, and that is through the various regulatory powers of the LVT. That is a slightly different subject and too big a topic to try to cover within this book. It is treated in some detail by the previous volume *Right to Manage and Service Charges: The New Regime*. Information can also be obtained from (among other sources) the LVT's governing body, the Residential Property Tribunals Service, "RPTS", or the Leasehold Advisory Service, "LEASE". Details of these organisations can be found in the useful addresses section at the back of this book.

As an aide-memoire, the list of applications which can be made to the LVT against landlords and management companies includes:

- appointment of a manager (the grounds for which include breach of covenant by the landlord)
- scrutiny of service charges
- reasonableness of insurance premiums
- scrutiny of administration charges.

Additionally, leaseholders may exercise the right to manage or the right to enfranchise without having to prove any fault on the landlord's part. Moreover, when sections 152, 153 and 156 of the Commonhold and Leasehold Reform Act 2002 eventually come into force, leaseholders will have a statutory right to withhold service charge payments as a weapon to make landlords comply with accounting requirements.

13 Costs and Administration Charges: A Summary

The question of the costs of enforcement has been addressed throughout these pages as it has come up in the context of a particular chapter. Costs may be payable contractually under a covenant, as part of the terms for relief from forfeiture, as costs of the action awarded by a court or the LVT or an arbitrator, or through the service charge. Different rules, statutes, case law and means of assessment apply in each context, but there is one common thread: full costs recovery is becoming increasingly difficult.

It is inevitable that costs will be incurred in any enforcement case. Whether it be thousands of pounds in legal fees or in terms of the valuable time of a property manager, some expenditure of money and/or time is unavoidable. The key is to minimise the irrecoverable element by a combination of:

- keeping the overall outlay down by using the most economic approach for each case and relying where possible on in-house staff
- concentrating on cases which can be won
- avoiding having to take enforcement measures in the first place
- avoiding costs penalties by mastering the appropriate procedures and selecting the suitable forum for the particular case (ie court, LVT or arbitration)
- being aware of strong arguments for costs submissions.

To a large extent, all these factors have been present for a considerable time, although the pressure on the scale of costs awarded by the court has been stepped up by the advent of the CPR 1998. There are two more recent changes which look likely to have a depressing effect on costs recovery, courtesy of the Commonhold and Leasehold Reform Act 2002, "the 2002 Act": the question mark over the whole area of recoverability of costs before the LVT and the new rules and regulations concerning administration charges.

The purpose of this chapter is to give an overview of the current position on costs recovery following the impact of the 2002 Act, with special reference to the decisions of the LVT, and the way in which administration charges fit in with all that. Not all the LVT decisions mentioned were made in enforcement cases, but they all give some clue as to the thought processes which the Tribunals are likely to apply.

(Where LVT cases are cited below, the reference given is the number allocated to it on the website of the Leasehold Advisory Service *www.lease-advice.org.uk* as this remains by far the simplest way to access LVT decisions. They are not official citations and should not be quoted in that way if used in arguments or submissions. The official citations can be found on each decision when accessed through the LEASE website.)

Costs of LVT proceedings

The LVT's power to award costs

(See Chapter 3 for a description of the Tribunal's powers on costs.)

Under paragraph 10 of schedule 12 to the 2002 Act, the LVT now has power to award costs as between the parties before it, but only to a limited extent. In *White* v *Younger* (LEASE Ref: 1351), costs of £500 (the maximum permissible) were awarded to the respondent leaseholder, the landlord having unreasonably prolonged the proceedings notwithstanding the tribunal's notification in directions that the landlord's case was fatally flawed.

In *GB Investments Ltd* v *Kavur* (Lease Ref: 1359) costs of £250 plus reimbursement of the application fee were awarded to the applicant landlord where the respondent had insisted on an oral hearing but failed to attend, giving no justifiable reason, and failing to act promptly or give full replies to the LVT's subsequent inquiries.

The LVT has the power to prevent a landlord passing on his costs of an LVT case to the service charge under section 20C of the Landlord and Tenant Act 1985 (see below). In *Walker* v *Grace Ltd* (LEASE Ref: 1273) the tribunal held that it did not need to make such an order if it considered that such costs were irrecoverable anyway. In this case, following (in its view) *St Mary's Mansions Ltd* v *Limegate Investment Co Ltd* [2003] E&CR, costs of defending a leaseholder's action under section 27A of the 1985 Act (to determine whether service charges were payable) were not covered by service charge provisions in the lease which included:

> The reasonable charges and expenses of the Lessor or the fees of the managing agents employed by the Lessor for the general management of the Building on behalf of the Lessor and/or the collection of the rents (including service charges) of the flats thereon.

or

> Legal or other costs incurred by the Lessor for the purposes of enforcing compliance by any of the owners lessees or occupiers for the time being of any of the flats comprised in the Building of any of the covenants similar to those contained in the Fifth Schedule hereto.

(Apparently, all of the covenants listed in the fifth schedule were related to nuisance issues.)

This tribunal also followed the reasoning of another tribunal in *Stoker* v *Urbanpoint Property Management Ltd* (LEASE Ref: 1112) which took the view that by paragraph 10(4) of schedule 12 to the Commonhold and Leasehold Reform Act 2002 "the recovery of such costs from another person is now absolutely precluded by statute".

Paragraph 10(4) of schedule 12 provides that no-one can be required to pay another's costs incurred in LVT proceedings except by determination by the LVT within its limited powers. This would seem to suggest that contractual provisions for costs (such as a covenant in a lease covering the costs of enforcement or forfeiture) will be of no effect in so far as costs at the LVT are concerned.

It has been suggested (including by the LVT itself — see for example *LB Sutton* v *Leaseholders of Benhill Estate* (LEASE Ref: 1239)) that the exclusion of recovery of costs incurred before the LVT except through the LVT's own limited power to award costs is restricted to the LVT being unable to order further recovery. In other words, the parties' contractual rights and remedies are unfettered except in enforcement cases through the LVT, and a landlord could still sue for such costs as a debt. If he attempted to forfeit on that ground however, the case is quite likely to come before the LVT. Could the LVT determine a breach in these circumstances? Possibly, but it seems a little illogical. This is the sort of unresolved issue inevitable at this early stage of development. Sooner or later a decision of the Lands Tribunal or the Court of Appeal will give firmer guidance, no doubt.

Costs of court proceedings

The various chapters dealing with debt recovery, declarations, forfeiture and possession proceedings cover the courts' treatment of costs issues within those respective types of cases, but what of the costs of court proceedings following a transfer to the LVT?

In *Imperial House Management Co Ltd* v *Howard* (LEASE Ref: 1294) the tribunal disposed of an application by the landlord and awarded "the Landlord's costs of the Court proceedings at £200". There is no discussion in the decision of how the LVT's authority to award and assess the costs of court proceedings arises. The tribunal went on to direct that "Any other costs may be recovered from the tenant in the form of service charges where provided for in the lease".

The LVT has no statutory authority to assess the costs of court proceedings, but it does have the authority to determine to what extent administration charges are payable. The question of administration charges is discussed below, but it would appear there is a possible overlap here which could easily cause confusion. Once again, this is the type of point which may be clarified in an appeal case in due course.

Costs recovery through the service charge

Section 20C of the Landlord and Tenant Act 1985 gives the LVT the power to order in appropriate circumstances that a landlord may not recover the costs of proceedings from the service charge (as opposed to from individual defaulting leaseholders).

In *Tenants of Langford Court* v *Doren Ltd*, a Lands Tribunal case of 2001, Judge Rich said:

> Where, as in the case of the LVT [at that time], there is no power to award costs, there is no automatic expectation of an order under section 20C in favour of a successful tenant, although a landlord who has behaved improperly or unreasonably cannot normally expect to recover his costs of defending such conduct, in my judgement the primary consideration that the LVT should keep in mind is that the power to make an order under

> section 20C should be used only in order to ensure that the right to claim costs as part of the service charge is not used in circumstances that makes its use unjust.

In the same judgment:

> Oppressive and, even more, unreasonable behaviour however is not found solely amongst landlords. Section 20C is a power to deprive a landlord of a property right. If the landlord has abused his rights or used them oppressively that is a salutary power, which may be used with justice and equity; but those entrusted with the discretion given by section 20C should be cautious to ensure that it is not itself turned into an instrument of oppression.

In *Ash* v *Ustunsurmeli* (LEASE Ref: 1374) (in what was, generally, an unusual decision) the Tribunal dismissed an application by the applicant leaseholders to exclude the landlord's costs of the proceedings from the service charge, but restricted the amount recoverable to £2,500 including VAT and disbursements (half the amount sought by the landlord).

In *Lewis* v *Warden Housing Association* (LEASE Ref: 1367) the LVT awarded the applicant leaseholder his application fee and made an order under section 20C of the Landlord and Tenant Act 1985. The tribunal found that the landlord housing association had not done enough to give the leaseholder sufficient information to prevent the necessity of the proceedings and, in particular in a sheltered development,

> managers of such housing had a responsibility, amongst others, to ensure that changes in service charges were reasonable and clearly explained to residents.

In *O'Callaghan* v *Trustees of R J Birchley (deceased)* (LEASE Ref: 1378) the tribunal made an order under section 20C in favour of the applicant leaseholders notwithstanding that they had had only limited success and the landlord had already offered to reduce the amount charged.

Recoverability of costs

(In any event, landlords may only recover costs through the service charge if the lease authorises such expenditure as a service charge item.)

In *Carberry* v *Mayor and Burgesses of the Royal Borough of Kensington and Chelsea* (LEASE Ref: 1381) the tribunal relied on *Sella House Ltd* v *Mears* [1989] 12 EG 67 and *Gilje* v *Charlesgrove Securities Ltd* [2001] EWCA 1777 in determining that: "The cost of management of the building ... and the Lessors' general administration expenses" could not be construed as allowing the landlord to charge its costs of the proceedings to the service charge.

Almost directly in opposition is the case of *Barros* v *Four Ennismore Ltd* (LEASE Ref: 1297) which held that LVT costs could be put to the service charge under the following wording (those words which the LVT felt were operative are highlighted):

> all legal costs incurred by the lessor (a) in the running and management of the building in the enforcement of the covenants conditions and regulations contained in the leases ... and (b) in making such applications and *representations and taking such action as the lessor shall reasonably think necessary in respect of any notice or order ... served under any statute* or order regulation or bye-law on the tenant or any under lessee of the flat[s] ... or on the lessor *in respect of the building ... or all or any of the flats therein.*

The LVT reached this decision in the absence of any submissions by the landlord and having been referred to the *St Mary's Mansions* and *Sella House* cases.

In *London Borough of Sutton* v *Leaseholders of the Benhill Estate* (LEASE Ref: 1435) the tribunal had been referred to both the *Langford Court* and *St Mary's Mansions* cases, they found that a very brief and general wording "to cover the cost and expenses incurred by the Lessor in carrying out the obligations or functions ..." including "To manage and conduct the management of the Estate and Building in a proper manner" was sufficient to give the landlord (a social landlord, which appears to have been significant in the eyes of the tribunal) authority in principle to pass the costs of proceedings to determine service charges on to the service charge payers.

The tribunal commented

> It is not possible to manage leasehold property without collecting the service charge funds to do so, and it is very difficult, if not impossible, under the present statutory regime to collect service charges where the liability to pay them is disputed without the ability to have recourse to the tribunal where the circumstances require it. It is in the interests of leaseholders as well as landlords that the funds should be available from service charges, reasonably incurred, to maintain buildings and provide

> services. Where a landlord applies precipitately or otherwise acts unreasonably in relation to the proceedings, the tenants can be protected by an order under section 20C. Where a landlord's costs in connection with the proceedings are excessive or disproportionate, they can be reduced by an order under section 27A on the ground that some of the costs were not reasonably incurred. We therefore conclude that the plain meaning of ... these leases, against the factual matrix within which it must be construed (including the fact that this is a social landlord), is that the landlord may recover its reasonable costs of coming to the tribunal as a necessary part of managing the Estate in a proper manner, subject to the tribunal's power to make an order under section 20C and subject to the reasonableness in amount of those costs.

These wise words are quoted at length as they may be of considerable interest and use to landlords, especially perhaps (following their logic) to social landlords and residents' management companies.

Similarly, in *Sinclair Gardens Investments (Kensington) Ltd* v *Avon Estates (London) Ltd* (LEASE Ref: 1393) the tribunal held that the phrase "in connection with or for the purpose of or in relation to the estate and the Block or any part thereof" allowed the landlord to recover from the service charge the costs of solicitors in representing the landlord on an application to appoint a manager (subsequently withdrawn).

Administration charges

Schedule 11 of the Commonhold and Leasehold Reform Act 2002 defines administration charges as follows:

> 1(1) In this Part of this Schedule "administration charge" means an amount payable by a tenant of a dwelling as part of or in addition to the rent which is payable, directly or indirectly —
> (a) for or in connection with the grant of approvals under his lease, or applications for such approvals,
> (b) for or in connection with the provision of information or documents by or on behalf of the landlord or a person who is party to his lease otherwise than as landlord or tenant,
> (c) in respect of a failure by the tenant to make a payment by the due date to the landlord or a person who is party to his lease otherwise than as landlord or tenant, or
> (d) in connection with a breach (or alleged breach) of a covenant or condition in his lease.

The definition is not wholly clear in terms of all its effects, but much of its meaning can be divined because of its similarity to the rather older statutory definition of a service charge (section 18, Landlord and Tenant Act 1985). For example, we can be confident that administration charges relate to charges which a leaseholder has covenanted to pay as opposed to items which are not contractually payable. What is not so obvious is whether legal and other costs payable under a covenant to pay costs of, say, a section 146 notice are intended to be covered. It is probably best to assume that they are unless and until there has been judicial clarification.

There was not a great deal of guidance available on this area by the time of writing. Without any great argument on the point, Section 146 costs were ruled out in *Clements* v *Pier Management Ltd* (LEASE Ref: 1354) where the costs were incurred in pursuance of sums which were not themselves validly due under the leases.

If it is correct that legal costs relating to enforcement are "administration charges", then solicitors' costs, counsel's fees and surveyors' charges incurred in an enforcement setting are liable to be assessed by the LVT if a leaseholder applies. The same would apply to any "late payment fees" or any other kind of default or penalty charge added to a leaseholder's account by the managing agent. The LVT will be asked to determine whether the charges are reasonable and whether they are payable under the terms of the lease. If they are not payable under the lease or under any order of the court, LVT or arbitrator then they are simply not recoverable. The same goes for interest charges.

As mentioned above, there is scope for conflict here between the powers of the LVT and the courts. For example, the court has the power to determine any costs of proceedings before it. It is possible that the court could assess the costs of an enforcement case, say for arrears. Those costs are then added to the leaseholder's account. What if the leaseholder then applies to the LVT to assess their reasonableness on the basis that they constitute an administration charge? Can the tribunal effectively re-open a matter already decided by the court? After all the LVT's standards of what is reasonable may easily differ. Hopefully a logical and equable solution will be found for this quandary.

Finally, any demand for administration charges must be accompanied by a summary of the leaseholder's rights and obligations as regards such charges. While property managers can be relied upon to do this with their own fees, it appears that they will have to remember to do so with any charges which arise from other sources, including the likes of solicitors', barristers' and surveyors' fees. This only arises of course if the fees are indeed "administration charges" which the leaseholder is being called upon to pay under his covenants.

Property managers may need to consider their procedures carefully in the light of these matters.

Conclusion

As can be seen from the LVT cases above, it is difficult to spot any trends or to predict how an individual tribunal is likely to decide a costs question in any particular case. There are enough cases where the tribunal has resisted the temptation to be too restrictive in its interpretations of leases to indicate that not all is lost — especially perhaps if the landlord in the case is a residents' management company or a social landlord. Nevertheless, any landlord or management company is still taking a substantial risk that a case before a court or the LVT is going to end up costing more in irrecoverable costs than the arrears in question. This is particularly so if there is any doubt that the wording of the lease covenants give clear authority for the charges to be made.

14 Conclusion

There have been some consistent threads running through this book, from which certain conclusions can be drawn. However it must be emphasised that there is no easy panacea to make enforcement cases simple or to ensure quiet, uncomplicated lives for those who manage leasehold property. Nevertheless, there are steps which can be taken which might help to protect against the burden being too great.

Horses for courses

Many practitioners fall into habits and routines, especially in the field of debt recovery procedures. This is natural, but it is a temptation which should be avoided. It is wise to remember that the law changes from time to time, as do the rules for putting the law into effect. Also, different categories of case respond best to different treatments — indeed each individual case ideally should receive full consideration of the most suitable strategy.

There is a very wide range of covenants in leases. A case could be made out for dealing with each of them differently; however, from the point of view of typical enforcement questions, they could be divided into four groups:

- ground rent
- service charges
- covenants of a technical nature

- covenants which impact directly upon other leaseholders (which is not to say that none of the above do, but some issues arouse more passion than others).

Ground rent

Undoubtedly forfeiture will occur less frequently for non-payment of ground rent on its own, both because of the new notification requirements and the restrictions on small amounts and short-term arrears. If the landlord can afford to wait however, forfeiture is still relatively straightforward in this context. The other option — debt recovery — is of doubtful use for small arrears anyway, although there is little risk of a defence in a rent case.

For the time being, most mortgage lenders will continue to discharge their borrowers' rent arrears, so an approach to them remains the obvious first step.

Service charges

Restrictions on forfeiture for service charges were a primary target of the Housing Act 1996. Although the fire of the Commonhold and Leasehold Reform Act 2002 was more widespread, there were still more restraints on service charge cases. The biggest difference is that a section 146 notice must now await a determination of the service charge rather than the other way around. This can be expected to have a substantial impact on costs as well as causing additional delays to cash flow.

So is it now worth even considering forfeiture for service charges (other than when the breach of covenant is admitted)?

Debt recovery through the county court is certainly simpler than forfeiture and, in general, it will be much quicker to obtain a final judgment. That is not the end of the matter though. The matter is thoroughly examined in Chapter 6, however it is worth noting a few points.

- Defended claims for under £5,000 are allocated to the small claims track in the county court, so very little is recoverable in costs. Most service charge claims will be for less than £5,000.
- There is a greater chance of recovering costs in an action for a declaration, but also a greater risk that the case will be transferred to the LVT, thus extending the life of the proceedings.

- The court can only deal with one leaseholder at a time whereas the LVT can determine the service charge for an entire estate, leading to economies of scale.
- Some (but not all) mortgage lenders will accept a county court judgment as sufficient evidence that the service charge is determined. This practice is likely to continue to decline.
- A defended debt claim can be transferred to the LVT, although it is less likely.
- Forfeiture is a long and difficult road, but it will nearly always bring in the cash. A creditor under a county court judgment may be forced to accept long term instalments or even possibly under-recovery.

The property manager is committed early to a choice of venue, but he should be looking ahead to the conclusion of the case and how best to bring in the cash. This means that no decision should be made until thought has been given to the most suitable type of enforcement of a judgment in the circumstances of the individual leaseholder. For example, if there is enough information to show that any of the standard enforcement measures available for a county court judgment (as described in Appendix 4) will work particularly well then it is obviously best to follow that approach. If, on the other hand, enough is known to indicate that no successful outcome awaits any of those facilities, then there is little point in even attempting a debt action — at least forfeiture will bring in the goods sooner or later.

Another early consideration is the strength of the covenants. A property manager in any doubt over whether the lease is likely to be interpreted in favour of the service charges he is seeking will be best advised to try for a default judgment or just leaving the claim to another day (probably when the flat is being sold) rather than risk close scrutiny by the court or LVT in an application for determination.

Covenants of a technical nature

This classification includes such things as registration of assignments or subletting, completing licences to assign, dealing appropriately with shares in the management company or insuring through the landlord's nominee.

There is no realistic formal alternative to forfeiture for breaches of these covenants, but there is the practical approach of just being awkward and uncooperative until breaches are remedied. There is nothing wrong in principle with obduracy so long as it does not extend to putting the landlord or management company in breach of covenant or some statutory or contractual obligation. In practice, this sort of problem is often best resolved by the exercise of tact and patience.

If there is no alternative to the ultimate sanction of forfeiture, it is best to combine it with other issues if possible. The courts do not look kindly on forfeiture being used for what they see as trivial or pedantic claims. Costs may be difficult to recover, and those costs could be substantial if two sets of proceedings were necessary (an action for a declaration followed by possession proceedings).

Covenants which impact directly upon other leaseholders

In nearly all conceivable circumstances, issues surrounding repairs or nuisance or the way the premises are used have a greater effect upon residents than they do upon landlords. When it comes to enforcing covenants of this nature, it can be assumed generally that the landlord is acting in the best interests of his leaseholders; either for the majority or for particularly vulnerable residents. This brings certain challenges and opportunities for the landlord and the property manager.

The most obvious opportunity is that, by working together with leaseholders in close harmony, goodwill can be enhanced enormously, with spin-offs in future co-operation and (put in the most mercenary way) cash flow. A happy leaseholder will contribute more readily than a disgruntled one.

The principal challenge is that cases of this type take a tremendous expenditure of time, effort and funds. Studies of the sort of case can be found at Chapters 8, 9 and 10 (and, to a certain extent, Chapter 12), but to recap some of the main considerations:

- forfeiture is the best long-term remedy even now, but there are interim measures
- injunctions and specific performance are particularly suitable for this type of situation
- the individually affected leaseholders can also seek injunctions and (in limited cases) press the authorities to seek Anti-Social Behaviour Orders or use other relevant powers (see Chapter 10)
- Evidence can be difficult to compile and witnesses are often reluctant to attend court — even those who first complained. Landlords and managers must tread carefully.

- costs are heavy in this type of case, but the landlord can expect more sympathy than usual
- landlords are usually entitled to require indemnities as to their costs from the complaining leaseholders. In the current climate it is advisable to take up that right — especially for residents' management companies
- there are special rules in repairs cases (see Chapter 8)
- other forms of dispute resolution such as mediation or conciliation may be appropriate.

Overall, for enforcement of leaseholders' covenants, it is clear that forfeiture is not dead yet, but the degree to which it has been weakened means that greater consideration should be given to the alternatives. None of the alternatives is perfect. They all have strengths and weaknesses and impose taxing requirements.

Reduce the need for enforcement

At the risk of sounding trite, it does no harm to remember that only covenants which have been broken need to be enforced. There are things which can be done to reduce the probability of breaches of covenant, for example:

- maximise goodwill through consultation, co-operation and transparency: these are required by best practice and legislation, but there is always room for improvement
- avoid arrears as far as possible through prudent budgeting, building reserves and credit management systems
- explain procedures and follow them consistently
- ensure that all parties fully understand both their rights and their responsibilities
- generally, encourage communication with and between leaseholders.

Reduce delay

When enforcement proves necessary, the quicker it can be completed the better for all concerned. The variety and complexity of the procedural requirements can seem daunting, especially since the implementation of the Commonhold and Leasehold Reform Act 2002, but there is nothing magical about them. Mastering the most efficient use of the procedures of the court and the LVT should pay for itself in terms of improved cash flow. (The capability to conduct enforcement actions comparatively painlessly will also represent a considerable selling point for property managers, whether they conduct the cases themselves or employ others.)

Minimise costs

The biggest single issue in litigation of any type is always going to be how much it costs. No form of proceedings or dispute resolution will ever be free of costs and the days of complete indemnities in forfeiture cases are long gone. The question is how to keep costs to a minimum, and particularly irrecoverable costs.

The first point is that each of the strategies outlined earlier in this chapter will have the effect of cutting costs: the more efficiently the case is run, the less it will cost.

The second key issue is to cut unnecessary expenditure. In a few cases, legal representation is practically unavoidable — injunctions and specific performance claims are the most obvious among common enforcement actions. Otherwise, most property managers are perfectly capable of conducting the various procedures and proceedings discussed in this book. If legal advice and assistance is needed from time to time during the course of a case, then by all means it should be obtained, but where matters can be progressed "in house" the opportunity should be grasped.

Training

Training opportunities abound currently for property managers, and this is especially apparent with the possibility of achieving specific professional qualifications and accreditation with the Institute of Residential Property Management (IRPM). There are also increasing numbers of courses for directors and officers of residents' management companies together with very helpful publications from organisations such as the Federation of Private Residents' Associations (FPRA) and LEASE.

Taking advantage of training and continuing to take advantage of it has to be the way forward in the enforcement context as in everything else. Training to conduct enforcement proceedings is only part of that. Learning how to improve property management skills generally can only assist in reducing to a minimum the scope for disputes and dissatisfaction, and thus the need to exercise enforcement powers.

Understanding the lease

Everything in leasehold matters springs from the lease: what it says and what it means. Statute will modify its effects but it will not alter its meaning fundamentally and neither will it insert things which are not there (not for the landlord's benefit anyway). A lease without a forfeiture clause cannot be forfeited. A lease which does not provide for costs recovery means that no costs can be recovered under it (unless costs can be recovered in other ways, such as the costs of court proceedings). If a lease does not provide for interest, it cannot be added willy-nilly to a leaseholder's account.

Finally, an ability to understand and apply the law is a wonderful thing, but the one absolute necessity which supersedes even that is a thorough knowledge and understanding of the leases with which the manager works.

The future

One thing we can predict with absolute certainty is that more legislation will come in due course. Eventually, forfeiture will be abolished, but first something else will need to be developed to replace it. The Law Commission has been working on ideas for around 20 years to date. Until then, the picture will change in fits and starts as ever.

The Commonhold and Leasehold Reform Act 2002 is still relatively new, especially as it has affected enforcement of covenants. Case law will be brought to bear over the next couple of years which will refine and clarify the legislation, helping us to comprehend it better and thus to apply it with a steadier touch.

It also remains to be seen how much difference (if any) will be made by commonhold.

As ever, the circumstances surrounding residential leasehold property are fraught with interest.

Appendix 1

Case Study: Background

The locus in quo

Property:	Tumbledown Mansions, Dystopia Road, Backwater, Nossex
Built:	Purpose-built 1985
Number of flats:	20, all on 99 year leases from 1985

Dramatis personae

Freeholder:	Reston Laurels Plc
Management company:	Tumbledown Management Limited ("TML")
Managing agent:	Reddy & Willing (formerly Reddy, Willing & Abel, but Abel left some time ago)
TML's solicitors:	Swift & Sharp

Ground floor

Flat 1:	Mr Walter Martinet, TML director, a semi-retired accountant
Flat 2:	Mrs Woolley, a widow
Flat 3:	Mr Hickok and Miss Oakley, tenants in common, (Mr Hickok's uncle is a builder)

Flat 4:	Retained by Reston Laurels, let out on an assured shorthold tenancy to Mr R Van Winkle, an elderly reclusive gentleman
Flat 5:	Mr A Wohl, a buy-to-let investor, resident abroad, billing address c/o "Play A Let" letting agents: subtenants Mr and Mrs T Dance (an elderly couple)

First floor

Flat 6:	Mr A Wohl: subtenant Ms Q Overend (works at a local pharmacy)
Flat 7:	Miss V Keen, a trainee solicitor
Flat 8:	Mr and Mrs P Witt, schoolteachers, Mr Witt is a director of TML
Flat 9:	Mr and Mrs P Nutt, early retired
Flat 10:	Mr Lars Straw, works at a gym

Second floor

Flat 11:	Mark Bench, a cartographer, director of TM
Flat 12:	Barry Chambers and Phyllis Potts, Barry is unemployed and Phyllis is a part-time cook
Flat 12A:	Mr and Mrs Hoodoo (Lou and Sue), an accident prone young couple
Flat 14:	Doug Root and Fallon Branch, landscape gardeners, Fallon is an RMC director
Flat 15:	Ms G String, a single lady

Third floor

Flat 16:	Mr A Wohl: subtenants Mr and Mrs R Gument
Flat 17:	Mr Owen Munny, constantly in arrears with his service charge contributions
Flat 18:	Holly and Ivy Andrews, sisters and interior designers
Flat 19:	Miss M Barras, a shy civil servant
Flat 20:	Mr and Mrs Rockett, an early retired couple, Mr Rockett has a short temper and complains of noise from the flat below (15)

Appendix 2
Possession Claim

On the following pages a sample of a Claim Form for possession of property and Particulars of claim for possession are reproduced. These can be downloaded from *www.hmcourts-service.gov.uk*

Claim form for possession of property

In the	Backwater County Court
Claim No.	

Claimant
(name(s) and address(es))

Reston Laurels Plc
Freeholder House
West Street
Backwater
Nossex

SEAL

Defendant(s)
(name(s) and address(es))

Mr A Wohl
Flat 5, Tumbledown Mansions
Dystopia Road
Backwater
Nossex

The claimant is claiming possession of :

Flat 5, Tumbledown Mansions
Dystopia Road
Backwater
Nossex

which (includes) ~~(does not include)~~ residential property. Full particulars of the claim are attached.
(The claimant is also making a claim for money).

This claim will be heard on: 20 at am/pm

at

At the hearing

- The court will consider whether or not you must leave the property and, if so, when.
- It will take into account information the claimant provides and any you provide.

What you should do

- Get help and advice immediately from a solicitor or an advice agency.
- Help yourself and the court by **filling in the defence form** and **coming to the hearing** to make sure the court knows all the facts.

Defendant's name and address for service

Mr A Wohl
Flat 5, Tumbledown Mansions
Dystopia Road
Backwater
Nossex

Court fee	£ 150.00
Solicitor's costs	£ TBA
Total amount	£ TBA

Issue date	

N5 Claim form for possession of property (06.04)

Printed on behalf of The Court Service

Claim No.

Grounds for possession

The claim for possession is made on the following ground(s):

- [x] rent arrears
- [ ] other breach of tenancy
- [x] forfeiture of the lease
- [ ] mortgage arrears
- [ ] other breach of the mortgage
- [ ] trespass
- [ ] other *(please specify)* ____________

Anti-social behaviour

The claimant is alleging:

- [ ] actual or threatened anti-social behaviour
- [ ] actual or threatened use of the property for unlawful purposes

Is the claimant claiming demotion of tenancy? [ ] Yes [x] No

See full details in the attached particulars of claim

Does, or will, the claim include any issues under the Human Rights Act 1998? [ ] Yes [x] No

Statement of Truth

*~~(I believe)~~(The claimant believes) that the facts stated in this claim form are true.
*I am duly authorised by the claimant to sign this statement.

signed ____________ date ____________

*~~(Claimant)(Litigation friend *(where the claimant is a child or a patient)*)~~(Claimant's solicitor)

**delete as appropriate*

Full name Raymond Zoroaster Sharp

Name of claimant's solicitor's firm Swift & Sharp

position or office held Partner
(if signing on behalf of firm or company)

Claimant's or claimant's solicitor's address to which documents or payments should be sent if different from overleaf.

Messrs Swift & Sharp
45 East Street
Backwater
Nossex

Ref: RZS

Postcode

if applicable	
Ref. no.	
fax no.	
DX no.	
e-mail	
Tel. no.	

Particulars of claim for possession
(rented residential premises)

Name of court Backwater County Court	Claim No.
Name of Claimant Reston Laurels Plc	
Name of Defendant Mr A Wohl	

1. The claimant has a right to possession of:
   Flat 5, Tumbledown Mansions
   Dystopia Road
   Backwater
   Nossex

2. To the best of the claimant's knowledge the following persons are in possession of the property:
   Mr and Mrs T Dance

About the tenancy

3. (a) The premises are let to the defendant(s) under a(~~n~~) lease for 99 years ~~tenancy~~ which began on 24th June 1985

   (b) The current rent is £100.00 and is payable each ~~(week) (fortnight) (month)~~. ~~(other~~ year)

   (c) Any unpaid rent or charge for use and occupation should be calculated at £0.27 per day.

4. The reason the claimant is asking for possession is:
   (a) because the defendant has not paid the rent due under the terms of the tenancy agreement.
   (Details are set out below)~~(Details are shown on the attached rent statement)~~
   The Defendant has failed to pay the rent due since 25th December 2000. The rent is payable half yearly in advance on 24th June and 25th December in each year. The arrears now amount to £400.00

   (b) because the defendant has failed to comply with other terms of the tenancy.
   Details are set out below.

   (c) because: (including any (other) statutory grounds)

N119 Particulars of claim for possession (rented residential premises) (06.04) *Printed on behalf of The Court Service*

5. The following steps have already been taken to recover any arrears:
Demands and reminders have been sent to the Defendant at the property and to his agents. No payment or other response has been received from the Defendant.

6. The appropriate (notice to quit) (notice of breach of lease) (notice seeking ~~possession) (notice seeking a demotion order) (*other*) was served on the defendant on 20 .~~

About the defendant

7. The following information is known about the defendant's circumstances:
The Defendant resides abroad. He owns the leasehold interest in three flats at Tumbledown Mansions including this property. All three flats are currently sub-let.

About the claimant

8. The claimant is asking the court to take the following financial or other information into account when making its decision whether or not to grant an order for possession:

Forfeiture

9. ~~(a) There is no underlessee or mortgagee entitled to claim relief against forfeiture.~~

or (b) Mr and Mrs T Dance of Flat 5, Tumbledown Mansions, Dystopia Road, Backwater Nossex are

~~is~~ entitled to claim relief against forfeiture as underlessee ~~or mortgagee.~~

What the court is being asked to do:

10. The claimant asks the court to order that the defendant(s):

(a) give the claimant possession of the premises;

(b) pay the unpaid rent and any charge for use and occupation up to the date an order is made;

(c) pay rent and any charge for use and occupation from the date of the order until the claimant recovers possession of the property;

(d) pay the claimant's costs of making this claim.

(e) see Rider

11. In the alternative to possession, is the claimant asking the court to make a demotion order?

☐ Yes ☑ No

Demotion claim

This section must be completed if the claim includes a claim for demotion of tenancy in the alternative to possession

12. The demotion claim is made under:

☐ section 82A(2) of the Housing Act 1985

☐ section 6A(2) of the Housing Act 1988

13. The claimant is a:

☐ local authority ☐ housing action trust ☐ registered social landlord

14. Has the claimant served on the tenant a statement of express terms of the tenancy which are to apply to the demoted tenancy?

☐ Yes ☐ No

If Yes, please give details:

15. The claimant is claiming demotion of tenancy because:
State details of the conduct alleged

Statement of Truth

*~~(I believe)~~(The claimant believes) that the facts stated in these particulars of claim are true.
* I am duly authorised by the claimant to sign this statement.

signed ______________________ date ______________________

~~(Claimant)(Litigation friend~~~~(where claimant is a child or a patient)~~*)(Claimant's solicitor)
**delete as appropriate*

Full name Raymond Zoroaster Sharp

Name of claimant's solicitor's firm Swift & Sharp

position or office held Partner
(if signing on behalf of firm or company)

Rider

Particulars of claim for possession

Paragraph 10

(e) Pay interest on the arrears pursuant to section 69 County Courts Act 1984 at the rate of 8% pa amounting to the date hereof to £xx.xx and continuing hereafter at a daily rate of £x.xx

Notes to completing these forms

As with all prescribed forms, they are subject to change. Practitioners are advised to carry out regular checks to ensure that their stocks are up to date. Electronic versions of prescribed forms will generally prompt insertions at the appropriate points; however, some points need to be deleted and others need to be typed in somehow, depending upon the available technology. Many of the deletions and additions will need to be made after the form has been printed, so it is important to remember to make these and to take copies of the final versions.

N5 claim form

This is a straightforward form, the successor to the old possession summons. Care needs to be taken to ensure that accurate names and addresses are inserted at each prompt. Although the amount of the court fee is fixed (and this also changes from time to time) the amount of solicitors' costs is not — unless the case is of the type for which tables of fixed fees have been laid down, and forfeiture does not come into that category. "TBA" in this situation means "to be assessed".

N119 Particulars of claim

As is suggested by its sub-title "(rented residential premises)" the N119 form is used for all types of residential tenancy and it has to be modified somewhat to be made relevant to long leases. To begin with, at paragraph 3(a), the word "tenancy" should be replaced by "lease" (although this is not crucial, since a lease is a form of tenancy). Indeed, it may make matters clearer if the insertion reads "lease for 99 years" (or whatever the term may be).

The answer to 3(c) is calculated simply by dividing the annual rent by 365. This becomes important later for determining what mesne profits are payable up to the date of recovering possession.

The details for paragraph 4 are relatively simple in a possession claim for non-payment of rent alone. For a case which involves any other breaches of covenant a rider will often be necessary to explain matters fully. It is unlikely in a long leasehold case that any particulars of statutory grounds will be required for 4(c).

Opinions are divided on the need to complete paragraph 5, but it is unlikely to be prejudicial to do so. A brief summary will be sufficient. The same applies to paragraph 7.

Paragraph 6 would need to be completed had a Notice been served under section 146 of the Law of Property Act 1925; however no such notice is needed in a rent case concerning a long lease.

It is rare for paragraph 8 to be completed, but it may be useful to do so if, for example, the landlord is a residents' management company which has been put at risk of insolvency by the defendant's default in payment.

Paragraph 9 is vital since this paragraph identifies any other parties who may be entitled to apply for relief from forfeiture (see Chapter 2). This information is needed by the court which is under a duty to serve copies of the proceedings upon anybody identified in this paragraph. In this case, Mr and Mrs Dance as subtenants are indicated, but a search at HM Land Registry has established no mortgage lenders or others with a charge on the flat.

The meat of the claim is contained in paragraph 10 and additions are necessary even in a simple rent claim; for service charges or other breaches far more substantial addenda would be needed. 10(c) is the claim for mesne profits (see above). 10(d) has a very bare claim to costs; if it is being argued that the claimant is entitled to contractual costs under the lease a rider should be added for the purpose.

In our scenario, a rider has been added as 10(e) to claim statutory interest under section 69 of the County Courts Act 1984. If interest is provided for in the lease it is usual to claim that instead; the award of interest is in the court's discretion but it is thought that it is more likely to be awarded under a contractual provision. There is an argument however that statutory interest is preferable at the moment because the rate is higher in the current climate than is achievable under many of the formulae contained in leases (2% above base for example).

When claiming interest on this form, the figures for interest accrued and a daily rate should be calculated and inserted.

Appendix 3

Application to the Leasehold Valuation Tribunal

On the following pages a Sample Residential Property Tribunal Service Form S27A is reproduced. This form can be downloaded from *www.rpts.gov.uk*

Application Form
S27A Landlord and Tenant Act 1985
Application for a determination of liability to pay service charges

This is the correct form to use if you want the Leasehold Valuation Tribunal to determine the liability to pay any service charge. This includes the question of whether or not the service charge is reasonable.

Please do not send any documents with this application form except a copy of the lease. If and when further evidence is needed you will be asked to send it in separately. If you have any questions about how to fill in this form or the procedures the Tribunal will use, please call us on 0845 600 3178. **Please send this completed application form together with the application fee and a copy of the lease to the appropriate panel (see page 6 for panel addresses).**

1. DETAILS OF APPLICANT(S) (if there are multiple Applicants please continue on a separate sheet)

Name SEE SEPARATE SHEET

Address (including postcode) ______

Address for correspondence *(if different)* ______

Telephone: *Day:* ______ *Evening:* ______ *Mobile:* ______

Email address: ______

Capacity (e.g. landlord/tenant/managing agent) ______

Representative details ______

Where details of a representative have been given, all correspondence and communications will be with them until the tribunal is notified that they are no longer acting.

2. ADDRESS (including postcode) OF PROPERTY (if not already given)

TUMBLEDOWN MANSIONS,
DYSTOPIA ROAD,
BACKWATER, NOSSEX

3. BRIEF DESCRIPTION OF PROPERTY *(e.g., 2-bedroom flat in Victorian block)*

RESIDENTIAL BLOCK OF 20 PURPOSE-BUILT FLATS, CONSTRUCTED IN 1985

1

4. DETAILS OF RESPONDENT(S) (if there are multiple Respondents please continue on a separate sheet)

Name SEE SEPARATE SHEET

Address (including postcode) ____________________

Address for correspondence *(if different)* ____________________

Telephone: *Day:* ________ *Evening:* ________ *Mobile:* ________

Email address *(if known)*: ____________________

Capacity *(e.g. landlord/tenant/managing agent)* ____________________

5. DETAILS OF LANDLORD (if not already given)

Name RESTON LAURELS PLC – SEE SECTION 1

Address (including postcode) ____________________

Telephone: *Day:* ________ *Evening:* ________ *Mobile:* ________

Email address *(if known)*: ____________________

6. DETAILS OF ANY RECOGNISED TENANTS' ASSOCIATION (if known)

Name of Secretary NONE

Address (including postcode) ____________________

Telephone: *Day:* ________ *Evening:* ________ *Mobile:* ________

Email address *(if known)*: ____________________

2

7. SERVICE CHARGES TO BE CONSIDERED BY THE TRIBUNAL

a. **Service charges for past years**

Please list years for which a determination is sought

1. ________________ 2. ________________

3. ________________ 4. ________________

5. ________________ 6. ________________

For each service charge year, fill in one of the sheets of paper entitled **Service Charges in Question.**

b. **Service charges for current or future years**

Please list years for which a determination is sought

1. ________________ 2. ________________

3. ________________ 4. ________________

5. ________________ 6. ________________

For each service charge year, fill in one of the sheets of paper entitled **Service Charges in Question.**

8. OTHER APPLICATIONS

Do you know of any other cases involving either: (a) the same or similar issues about the service charge as in this application; or (b) the same landlord or tenant or property as in this application ? If so please give details.

NO

9. LIMITATION OF COSTS

If you are a tenant, do you wish to make a s20C application *(see Guidance Notes)* YES ☐ NO ☐

If so, why? ________________

Guidance Notes
Some leases allow a landlord to recover costs incurred in connection with the proceedings before the LVT as part of the service charge. Section 20C of the Landlord and Tenant Act 1985 gives the tribunal power, on application by a tenant, to make an order preventing a landlord from taking this step. If you are a tenant you should indicate here whether you want the tribunal to consider making such an order.

3

10. CAN WE DEAL WITH YOUR APPLICATION WITHOUT A HEARING?

If the Tribunal thinks it is appropriate, and all the parties agree, it is possible for your application to be dealt with entirely on the basis of written representations and documents and without the need for parties to attend and make oral representations. This means you would not be liable for a hearing fee of £150 but it would also mean that you would not be able to explain your case in person. Please let us know if you would be happy for your application to be dealt with in this way.

I would be happy for the case to be dealt with on paper if the Tribunal thinks it is appropriate YES ☑ NO ☐

NB: Even if you have asked for a determination on paper the Tribunal may decide that a hearing is necessary. Please go on to answer questions 11 to 13 on the assumption that a hearing will be heard.

11. TRACK PREFERENCES

We need to decide whether to deal with the case on the Fast Track or the Standard Track. (see Guidance Notes for an explanation of what a track is). Please let us know which track you think appropriate for this case.

Fast Track ☐ Standard Track ☑

Is there any special reason for urgency in this case? YES ☐ NO ☐

If there is, please explain how urgent it is and why: ______________________

The Tribunal will normally deal with a case in one of three ways: on paper, on the "fast track", and on the "standard track." The fast track is designed for cases that need a hearing but are very simple and will not generate a great deal of paperwork or argument. A fast track case will usually be heard within 10 weeks of your application. You should indicate here if you think the case is very simple and can easily be dealt with. The standard track is designed for more complicated cases where there may be numerous issues to be decided or where, for example, a lot of documentation is involved. A standard track case may involve the parties being invited to a Pre-Trial Review which is a meeting at which the steps that need to be taken to bring the case to a final hearing can be discussed.

12. AVAILABILITY

If there are dates or days we must avoid during the next three months (either for your convenience or the convenience of any expert you may wish to call) please list them here.

Dates on which you will NOT be available ______________________

13. VENUE REQUIREMENTS

Please provide details of any special requirements you or anyone who will be coming with you may have (e.g., the use of a wheelchair and/or the presence of a translator) ______________________

In London cases are usually heard in Alfred Place, which is fully wheelchair accessible. Elsewhere hearings are held in local venues which are no all so accessible and the Clerks will find it useful to know if you or anyone you want to come to the hearing with you has any special requirements of this kind.

4

14. CHECK LIST

Please check that you have completed this form fully. The tribunal will not process your application until this has been done and it has both a copy of the lease and the application fee:

A copy of the lease(s) is/are enclosed ☑

A crossed cheque or postal order for the application fee (if applicable) is enclosed ☑

Amount of fee enclosed __________ Please put your name and address on the back of any cheque you send.

DO NOT send cash under any circumstances. Cash payment will not be accepted and any application accompanied by cash will be returned to the applicant.

Please ONLY send this application form, a copy of the lease and the application fee and nothing else.

Guidance notes:
The amount of the application fee will depend on the total amount of service charge that is in dispute. To find out how much you will need to pay you should consult the following table:

Amount of Service Charge in dispute	***Application Fee***
Not more than £500	*£50*
More than £500 but less than £1,000	*£70*
More than £1,000 but less than £5,000	*£100*
More than £5,000 but less than £15,000	*£200*
More than £15,000	*£350*

Fees should be paid by a crossed cheque made payable to or a postal order drawn in favour of the Office of the Deputy Prime Minister.

Waiver and Fees
You will ***not*** *be liable to pay a fee if you or your partner is in receipt of:*

- *Income Support*
- *Housing Benefit*
- *Income Based Job Seeker's Allowance*
- *A working tax credit, and either:*
  - *you or your partner receive a tax credit with a disability or severe disability element; or*
  - *you or your partner is also in receipt of a child tax credit*
- *a guarantee credit under the State Pensions Credit Act 2002*
- *certificate issued under the Funding Code which has not been revoked or discharged and which is in respect of the proceedings before the tribunal the whole or part of which have been transferred from the county court for determination by a tribunal*
- *A Working Tax Credit where the Gross Annual Income used to calculate the Tax Credit is £14,213 or less*

If you wish to claim a waiver of fees you must complete another form available from the Panel office. The waiver form will not be copied to other parties in the proceedings.

If you are making several applications at the same time, even if you are using different application forms or the applications relate to different parts of the Tribunal's jurisdiction, you do not have to pay a separate fee for each application. The overall fee will be the biggest of the fees payable for each application on its own.

If you are in any doubt about the amount of fee or have any other questions about how to fill in this form please telephone the RPTS help line on 0845 600 3178.

15. STATEMENT OF TRUTH

I believe that the facts stated in this application are true.

Signed: ____________________ Dated: __________

5

PANEL ADDRESSES

Northern Rent Assessment Panel
20th Floor, Sunley Tower, Piccadilly Plaza, Manchester M1 4BE

Telephone: 0845 1002614
Facsimile: 0161 2367 3656

Midland Rent Assessment Panel
2nd Floor, East Wing, Ladywood House
45-46 Stephenson Street
Birmingham B2 4DH

Telephone: 0845 1002615
Facsimile: 0121 643 7605

Eastern Rent Assessment Panel
Great Eastern House, Tenison Road, Cambridge CB1 2TR

Telephone: 0845 1002616
Facsimile: 01223 505116

London Rent Assessment Panel
2nd Floor, 10 Alfred Place, London WC1E 7LR

Telephone 020 7446 7700
Facsimile 020 7637 1250

Southern Rent Assessment Panel
1st Floor, 1 Market Avenue, Chichester PO19 1JU

Telephone: 0845 1002617
Facsimile: 01243 779389

6

SERVICE CHARGES IN QUESTION

PLEASE USE THE SPACE BELOW TO PROVIDE INFORMATION REGARDING EACH OF THE YEARS MENTIONED IN PART 7 OF THE MAIN APPLICATION FORM.

You will be given an opportunity later to give further details of your case and to supply the Tribunal with any documents that support it. At this stage you should give a clear outline of your case so that the Tribunal understands what your application is about.

The year in question ________________________________

A list of the items of service charge that are in issue (or relevant) and their value ________________________________

Description of the question(s) you wish the tribunal to decide ________________________________

Any further comments you may wish to make ________________________________

Rider

1. Details of applicants

Names (1) Reston Laurels plc and (2) Tumbledown Management Ltd

Address

(1) Freeholder House
West Street
Backwater
Nossex

(2) 45 East Street
Backwater
Nossex

Address for correspondence 69 North Street
Backwater
Nossex

Telephone [insert appropriate details]

Email address [insert appropriate details]

Capacity (1) Landlord and (2) management company

Representative details Ms Ethel Reddy, Reddy & Willing Property Managers, 69 North Street, Backwater, Nossex represents both Applicants

4. Details of respondents

Names

(1) Mr W B Hickok and Miss A Oakley
(2) Miss V Keen
(3) Mr Lars Straw
(4) Mr B Chambers and Miss P Potts
(5) Mr O Munny

Addresses

(1) Flat 3, Tumbledown Mansions, Dystopia Road, Backwater, Nossex
(2) Flat 7, Tumbledown Mansions, Dystopia Road, Backwater, Nossex
(3) Flat 10, Tumbledown Mansions, Dystopia Road, Backwater, Nossex
(4) Flat 12, Tumbledown Mansions, Dystopia Road, Backwater, Nossex
(5) Flat 17, Tumbledown Mansions, Dystopia Road, Backwater, Nossex

Addresses for correspondence	No other addresses are known to the applicants
Telephone	[insert appropriate details]
Email address	[insert appropriate details]
Capacity	Leaseholders

7. Service charges to be considered by the tribunal

[See notes below]

Notes to completing this form

1. The completion of paragraph 1 is not an exact science as the way in which it is completed will vary depending upon who is the applicant and who is the representative, and what is the representative's status. The important point is to ensure that the tribunal knows precisely the identity of the applicant and the appropriate contact details for the representative.
2. Obviously the full address including postcode should be inserted at all relevant points. Where the address of the subject property is concerned, it may be that specific flats should be quoted or simply the address of the block or estate. The choice should be dictated by applying common sense to the circumstances of the application.
3. The key points to be covered by the description of the property are whether it is a single flat or an estate and whether there is any commercial element. It is also useful to state whether the building is a conversion or purpose-built and a rough date of conversion or development.
4. Alternative addresses for respondents should be given if known. When confirming their capacity, few long leaseholders like to be known as "tenants" and the tribunal will understand that there is no distinction.
5. In the case of multiple parties, it is as well to confirm again the name of the landlord here to avoid any misunderstanding of who is who.
6. It is important to insert only the details of any recognised tenants' association within the meaning of section 29 of the Landlord and Tenant Act 1985, since only such an association has specific statutory rights.

7. The detail to be supplied for service charges to be considered will be crucial. In an application against multiple respondents such as this, it is essential to include the particulars for any years or future periods which form part of the claim against any one or more of the respondents. Note that this section needs to be completed together with the separate sheet for each year, one blank sheet of which is supplied as the last page of this downloadable form. If more than one sheet is required, it is wise to copy sufficient numbers of the blank sheet before inserting any details.
10. There are obvious advantages in speed and cost of electing to ask the tribunal to deal with the application on the basis of written representations only. Where the claim can be clearly understood and substantiated on the documents it is probably worth making this request, even where there are multiple parties. The claim can only be dealt with on this basis if all parties and the tribunal agree, and it is unlikely that the tribunal would agree if the respondents' replies disclose any complicated or especially contentious arguments.
11. Somewhat similar considerations apply to the selection of the appropriate track. The guidance notes on the printed form help to inform the decision. It is perhaps optimistic to request a fast track hearing with the number of respondents to this case, but the choice will still be influenced by the amount of the arrears and the length of time they cover: if the arrears consist of five contributions to the same account for example, the fast track would be more appropriate. This section also gives the opportunity to point out any urgency. If a case is urgent this should also be pointed out in the covering letter and followed up by telephone to the tribunal office.
12. It is wise to check dates to avoid with principal witnesses, directors of the applicant companies and their representatives at an early stage as it is not always easy to disturb a hearing date once it has been fixed.
14. The form prompts the applicant to check that the other necessary papers are submitted. The only documentation required at this stage is the lease. In this case either copies of the leases for each of the respondents' flats should be supplied or one of them with confirmation in the covering letter that all five leases are identical save for the flat number, execution dates and parties (assuming they are identical of course). This confirmation will carry more weight from a solicitor or licensed conveyancer. The fee to be paid

is determined by the amount in question: with five Respondents, the amount is the aggregate of their arrears. The form carries an explanatory table of fees.

15. See the notes on statements of truth under the heading "The Claim Form" in Chapter 5 (which deals with a county court claim).

Panel addresses: If you are in doubt over which region is appropriate, a telephone call to one or more panel office stating the postcode of the subject property should answer the point.

Appendix 4
Debt Claim

On the following pages a sample of a Claim Form is reproduced. This can be downloaded from *www.hmcourts-service.gov.uk*

Claim Form

In the	Backwater County **Court**
	for court use only
Claim No.	
Issue date	

SEAL

Claimant

Tumbledown Management Limited,
45 East Street,
Backwater,
Nossex

Defendant(s)

Holly Andrews (1) and Ivy Andrews (2),
Flat 18 Tumbledown Mansions,
Dystopia Road,
Backwater,
Nossex

Brief details of claim

Unpaid service charges due under a Lease dated 24th June 1985 of the residential property known as Flat 18, Tumbledown Mansions, Dystopia Road, Backwater, Nossex

Value

The Claimant expects to recover less than £5,000

Defendant's name and address

Holly Andrews (1) and Ivy Andrews (2),
Flat 18 Tumbledown Mansions,
Dystopia Road,
Backwater,
Nossex

	£
Amount claimed	2,000.00
Court fee	120.00
Solicitor's costs	N/A
Total amount	2,120.00

The court office at

is open between 10 am and 4 pm Monday to Friday. When corresponding with the court, please address forms or letters to the Court Manager and quote the claim number.

N1 Claim form (CPR Part 7) (01.02) *Printed on behalf of The Court Service*

Claim No.	

Does, or will, your claim include any issues under the Human Rights Act 1998? ☐ Yes ☑ No

Particulars of Claim ~~(attached)(to follow)~~

1. Under a lease dated 24th June 1985 between Reston Laurels Plc ("the Lessor") (1) Tumbledown Management Limited ("the Manager") (2) and Peter and Tinkerbell Pan ("the Lessees") the property known as Flat 18 Tumbledown Mansions, Dystopia Road, Backwater, Nossex ("the Property") was let to the Lessees. The lease was for a fixed term of 99 years commencing on 24th June 1985. A copy of the lease is annexed as Attachment A and it should be referred to for its full terms and effect.

2. On or about 13th April 1999 the Defendants acquired the Lessees' title to the Property. Annexed as Attachment B is a copy of the Office Copy Entries of the Defendants' leasehold title.

3. The Claimant's managing agents are Messrs Reddy & Willing.

4. The lease contains the following covenants on the part of the Lessees:
i) to pay rent;
ii) to pay a service charge at the times and in the manner set out in Clause 3 and the Fourth Schedule to the lease;
iii) to pay interest on any arrears of service charge caluculated in accordance with Clause 3(p).

5. The Defendants are in breach of the covenant set out in paragraph 4 ii) above, full details of which appear in the Schedule of Arrears annexed as attachment C.

And the Claimant claims against the Defendants:

a) The sum of £2,000 as detailed in the Schedule of Arrears
b) Interest pursuant to S.69 of the County Courts Act 1984 at 8% per annum amounting to the date hereof to £xx.xx and continuing hereafter at a daily rate of £xx.xx
c) Costs

Statement of Truth
*~~(I believe)~~(The Claimant believes) that the facts stated in these particulars of claim are true.
* I am duly authorised by the claimant to sign this statement

Full name Walter Martinet

~~Name of claimant's solicitor's firm~~

signed

*(Claimant)(Litigation friend)(Claimant's solicitor)

delete as appropriate

position or office held Director

(if signing on behalf of firm or company)

Messrs Reddy & Willing,
69 North Street,
Backwater,
Nossex

Claimant's or claimant's solicitor's address to which documents or payments should be sent if different from overleaf including (if appropriate) details of DX, fax or e-mail.

Notes to completing this form

As with all prescribed forms, it is subject to change. Practitioners are advised to carry out regular checks to ensure that their stocks are up to date. Electronic versions of prescribed forms will generally prompt insertions at the appropriate points; however, some points need to be deleted and others need to be typed in somehow, depending upon the available technology. Many of the deletions and additions will need to be made after the form has been printed, so it is important to remember to make these and to take copies of the final versions.

It is important to ensure that the parties' names and addresses are given accurately. Obviously post codes should be supplied when they are known. When there are multiple defendants, they should be identified as such, and obviously separate addresses given if they do not share an address.

The "Brief details of claim" should identify the areas of law concerned to enable the clerks to deal with the claim most efficaciously. Flagging "service charge" and "residential" in this section will be very helpful to the court. It is not necessary however to insert the particulars of the claim here as that information is required later.

The statement of value is of doubtful value in itself, although it will become relevant if there has to be a decision on allocation to a particular track (see below). The figure should not include interest or costs. It should be either the figure being claimed or state that claimant expects to recover no more than £5,000 or £15,000 or more than £15,000 (£5,000 and £15,000 being the notional limits for the small claims track and fast track respectively).

The court's issue fee is on a graduated scale, depending upon the amount claimed. The up-to-date figures can be ascertained from county court offices or the Courts Service *website www.hmcourts-service.gov.uk*. A figure for "solicitor's costs" should only be claimed if solicitors are acting. That is not to say that parties acting for themselves ("litigants in person") or in another representative capacity are not entitled to make a claim for costs within the proceedings, but the box on the form is to assist calculation of fixed costs which are only available to solicitors.

The particulars of claim can be contained in a separate document and attached to the claim form, but in a straightforward debt claim it should be possible to condense the particulars into the parameters of the form. The particulars should include:

- a concise statement of the facts relied upon
- reference to attached copies of any written agreements, contracts or other documents upon which the claim is based (the lease being the obvious document in a service charge case)
- if interest is to be claimed (see below), a statement to that effect
- details of the remedy sought by the Claimant.

Where interest is claimed, the particulars of claim should identify whether it is based on a contractual term or on statute and, if ascertainable, the figures should be calculated and a note given of the applicable dates for the calculations.

The example above refers to a separate schedule of arrears, but this is not necessary if the table required can be inserted into the space available on the form. One or two missed payments should be capable of insertion for example. Wherever it is placed, the schedule should briefly identify, describe and date each instalment. With service charges, the description should accord as closely as possible with the wording in the lease. For example, many older leases refer to "maintenance charge" rather than "service charge".

The form should be dated, and the date should be that to which the interest has been calculated.

The defence

IN THE BACKWATER COUNTY COURT

Claim No. [*As Claim Form*]

BETWEEN

TUMBLEDOWN MANAGEMENT LIMITED

Claimant

AND

HOLLY ANDREWS (1) AND IVY ANDREWS (2)

Defendants

DEFENCE

1. Paragraphs 1–4 of the particulars of claim are admitted.
2. It is denied that the defendants are in breach of covenant or that any sums as set out in the claimant's achedule of arrears are properly due to the claimant.

3. This is a claim for service charges and service charges are properly payable only to the extent that they are reasonable.
4. The service charges claimed by the claimant are unreasonable and excessive and are not due from the defendants unless and until they have been determined by the court or by the leasehold valuation tribunal.
5. The cost of the works of external redecoration included within the claimant's service charges are not due from the defendants as the works were carried out to a poor standard and the charges of the contractors were unreasonable and excessive.
6. The management fees of Reddy & Willing included within the claimant's charges are not due from the defendants as they have provided a poor service and their fees are unreasonable and excessive.
7. The defendants will abide by any determinations of service charge by the court or the leasehold valuation tribunal.
8. In the circumstances the defendants deny that the claimant is entitled to the sums claimed in the particulars of claim.

STATEMENT OF TRUTH

The defendants believe that the facts stated in this defence are true.

____

Holly Andrews

____

Ivy Andrews

Defendants

Dated ____________________

Notices about this case may be sent to the defendants at Flat 18 Tumbledown Mansions, Dystopia Road, Backwater, Nossex

Allocation questionnaire

On the following pages a sample of an Allocation Questionnaire is reproduced. This can be downloaded from *www.hmcourts-service.gov.uk*

Allocation questionnaire

To be completed by, or on behalf of,

Tumbledown Management Limited

who is ~~[1st][2nd][3rd][—]~~[Claimant]~~[Defendant]~~ ~~[Part 20 claimant]~~ in this claim

In the	Backwater County Court
Claim No.	
Last date for filing with court office	

Please read the notes on page five before completing the questionnaire.

You should note the date by which it must be returned and the name of the court it should be returned to since this may be different from the court where the proceedings were issued.

If you have settled this claim (or if you settle it on a future date) and do not need to have it heard or tried, you must let the court know immediately.

Have you sent a copy of this completed form to the other party(ies)? ☑ Yes ☐ No

A Settlement

Do you wish there to be a one month stay to attempt to settle the claim, either by informal discussion or by alternative dispute resolution? ☐ Yes ☑ No

B Location of trial

Is there any reason why your claim needs to be heard at a particular court? ☐ Yes ☑ No

If Yes, say which court and why?

C Pre-action protocols

If an approved pre-action protocol applies to this claim, complete **Part 1** only. If not, complete **Part 2** only. If you answer 'No' to the question in either Part 1 or 2, please explain the reasons why on a separate sheet and attach it to this questionnaire.

Part 1 **please say which protocol*

The* ______ protocol applies to this claim.

Have you complied with it? ☐ Yes ☐ No

Part 2

No pre-action protocol applies to this claim.

Have you exchanged information and/or documents (evidence) with the other party in order to assist in settling the claim? ☑ Yes ☐ No

N150 Allocation questionnaire (10.01) 1 *Printed on behalf of The Court Serv*

D Case management information

What amount of the claim is in dispute? £ 2,000.00

Applications

Have you made any application(s) in this claim? ☐ Yes ☑ No

If Yes, what for? (e.g. summary judgment, add another party) For hearing on

Witnesses

So far as you know at this stage, what witnesses of fact do you intend to call at the trial or final hearing including, if appropriate, yourself?

Witness name	Witness to which facts
Ms Ethel Reddy) Mr Walter Martinet)	The expenditure contained within the service charges claimed in these proceedings

Experts

Do you wish to use expert evidence at the trial or final hearing? ☐ Yes ☑ No

Have you already copied any experts' report(s) to the other party(ies)? ☐ None yet obtained ☐ Yes ☐ No

Do you consider the case suitable for a single joint expert in any field? ☐ Yes ☐ No

Please list any single joint experts you propose to use and any other experts you wish to rely on. Identify single joint experts with the initials 'SJ' after their name(s).

Expert's name	Field of expertise (eg. orthopaedic surgeon, surveyor, engineer)

Do you want your expert(s) to give evidence orally at the trial or final hearing? ☐ Yes ☐ No

If Yes, give the reasons why you think oral evidence is necessary:

2

continue over

Track

Which track do you consider is most suitable for your claim? Tick one box

☑ small claims track ☐ fast track ☐ multi-track

If you have indicated a track which would not be the normal track for the claim, please give brief reasons for your choice

E Trial or final hearing

How long do you estimate the trial or final hearing will take? ___ days 1 hours 30 minutes

Are there any days when you, an expert or an essential witness will not be able to attend court for the trial or final hearing? ☐ Yes ☑ No

If Yes, please give details

Name	Dates not available

F Proposed directions *(Parties should agree directions wherever possible)*

Have you attached a list of the directions you think appropriate for the management of the claim? ☐ Yes ☑ No

If Yes, have they been agreed with the other party(ies)? ☐ Yes ☐ No

G Costs

*Do **not** complete this section if you have suggested your case is suitable for the small claims track **or** you have suggested one of the other tracks and you do not have a solicitor acting for you.*

What is your estimate of your costs incurred to date? £

What do you estimate your overall costs are likely to be? £

In substantial cases these questions should be answered in compliance with CPR Part 43

3

Other information

Have you attached documents to this questionnaire? ☐ Yes ☑ No

Have you sent these documents to the other party(ies)? ☐ Yes ☐ No

If Yes, when did they receive them?

Do you intend to make any applications in the immediate future? ☑ Yes ☐ No

If Yes, what for? Summary Judgment

In the space below, set out any other information you consider will help the judge to manage the claim.

Signed Date [Date]

~~[Counsel][Solicitor][for the][1st][2nd][3rd][ ]~~
[Claimant]~~[Defendant][Part 20 claimant]~~

Please enter your firm's name, reference number and full postal address including (if appropriate) details of DX, fax or e-mail

[Address details]	if applicable	
	fax no.	
	DX no.	
Tel. no. Postcode	e-mail	

Your reference no.

4

Notes for completing an allocation questionnaire

- If the claim is not settled, a judge must allocate it to an appropriate case management track. To help the judge choose the most just and cost-effective track, you must now complete the attached questionnaire.
- If you fail to return the allocation questionnaire by the date given, the judge may make an order which leads to your claim or defence being struck out, or hold an allocation hearing. If there is an allocation hearing the judge may order any party who has not filed their questionnaire to pay, immediately, the costs of that hearing.
- Use a separate sheet if you need more space for your answers marking clearly which section the information refers to. You should write the claim number on it, and on any other documents you send with your allocation questionnaire. Please ensure they are firmly attached to it.
- The letters below refer to the sections of the questionnaire and tell you what information is needed.

A Settlement

If you think that you and the other party may be able to negotiate a settlement you should tick the 'Yes' box. The court may order a stay, whether or not all the other parties to the claim agree. You should still complete the rest of the questionnaire, even if you are requesting a stay. Where a stay is granted it will be for an initial period of one month. You may settle the claim either by informal discussion with the other party or by alternative dispute resolution (ADR). ADR covers a range of different processes which can help settle disputes. More information is available in the Legal Services Commission leaflet 'Alternatives to Court' free from the LSC leaflet line Phone: 0845 3000 343

B Location of trial

High Court cases are usually heard at the Royal Courts of Justice or certain Civil Trial Centres. Fast or multi-track trials may be dealt with at a Civil Trial Centre or at the court where the claim is proceeding. Small claim cases are usually heard at the court in which they are proceeding.

C Pre-action protocols

Before any claim is started, the court expects you to have exchanged information and documents relevant to the claim, to assist in settling it. For some types of claim e.g. personal injury, there are approved protocols that should have been followed.

D Case management information

Applications

It is important for the court to know if you have already made any applications in the claim, what they are for and when they will be heard. The outcome of the applications may affect the case management directions the court gives.

Witnesses

Remember to include yourself as a witness of fact, if you will be giving evidence.

Experts

Oral or written expert evidence will only be allowed at the trial or final hearing with the court's permission. The judge will decide what permission it seems appropriate to give when the claim is allocated to track. Permission in small claims track cases will only be given exceptionally.

Track

The basic guide by which claims are normally allocated to a track is the amount in dispute, although other factors such as the complexity of the case will also be considered. A leaflet available from the court office explains the limits in greater detail.

Small Claims track	Disputes valued at not more than £5,000 except · those including a claim for personal injuries worth over £1,000 and · those for housing disrepair where either the cost of repairs or other work exceeds £1,000 or any other claim for damages exceeds £1,000
Fast track	Disputes valued at more than £5,000 but not more than £15,000
Multi-track	Disputes over £15,000

E Trial or final hearing

You should enter only those dates when you, your expert(s) or essential witness(es) will not be able to attend court because of holiday or other commitments.

F Proposed directions

Attach the list of directions, if any, you believe will be appropriate to be given for the management of the claim. Agreed directions on fast and multi-track cases should be based on the forms of standard directions set out in the practice direction to CPR Part 28 and form PF52.

G Costs

Only complete this section if you are a solicitor and have suggested the claim is suitable for allocation to the fast or multi-track.

H Other Information

Answer the questions in this section. Decide if there is any other information you consider will help the judge to manage the claim. Give details in the space provided referring to any documents you have attached to support what you are saying.

5

Notes on completing this form

As with other prescribed court forms, many of the deletions and additions will need to be made after the form has been printed, so it is important to remember to make them and to take copies of the final versions. The official notes attached to the form are on the last page; nonetheless it is advisable to read them through first. The notes below are supplementary and are not intended as a substitute for the official guidance notes.

The parties are required to try to agree the contents of the allocation questionnaire. You should at least copy your draft to the other side and invite their agreement, but there is rarely time to do more than this in practice and still comply with the court's time limit for return of the completed form.

Defendants often request a one month stay without a great deal of thought, simply to buy time; similarly, many claimants make the request to overcome their surprise at an unexpected point in the defence. In a straightforward small claim there is usually little point in seeking a stay unless a negotiated settlement is realistically achievable.

Confusion can be caused by the references in part D to applications made and in part H to applications in the immediate future (and, for that matter, in part F to proposed directions). This is one of the reasons why the completion of the form should be planned in advance.

There is a case sometimes for requesting a different track than that indicated simply by the amount claimed. With a claim for less than £5,000 this will normally revolve around a complex point of law for which lengthy legal argument may be needed or the volume of evidence. If the request is made for such reasons, the multi-track will normally be the more appropriate. Just the fact that this is a service charge case and that service charge cases are notoriously complicated compared to other debt cases will not usually be regarded as adequate reason for allocation to other than the small claims track.

Estimating the length of a final hearing is never easy, especially at such an early stage in the proceedings. Any disputed case is likely to take at least an hour. Beyond that it is anyone's guess; none the less, it is important to make the attempt because there is a risk otherwise that the court will fix a minimal length of appointment which will result in a hearing which has to be adjourned "part heard" (unfinished). On the other hand, if too long an estimate is given, it could be a long time ahead before court time is found.

Most county courts will have preferred standard directions for most standard types of case. Unless you think there is anything particularly unusual for your small claim, it is as well to let the court use its own standard directions.

Some litigants use the section on costs with a view to intimidating their opponents even when (as in this case) there is no need to insert anything in this part of the form (part G). This practice should be avoided because it may be seen as an abuse of the process and lead to penalties — even possibly the claim being struck out.

In this case, the claimant has elected to apply for summary judgment. The application could be filed at the same time as the allocation questionnaire, in which case this should be referred to under "Other information". Indeed, PD 26 under the CPR states that an application should be made before or when filing the allocation questionnaire. It would also be appropriate to insert here, for example, that the same or similar issues as those raised in the defence have already been determined by the court elsewhere or by the LVT.

Application for Summary Judgment

On the following pages a sample of an Application Notice is reproduced. This can be downloaded from *www.hmcourts-service.gov.uk*

Application Notice

You should provide this information for listing the application

1. How do you wish to have your application dealt with
   - a) at a hearing? ☑ } *complete all questions below*
   - b) at a telephone conference? ☐
   - c) without a hearing? ☐ *complete Qs 5 and 6 below*
2. Give a time estimate for the hearing/conference ________(hours) 30 (mins)
3. Is this agreed by all parties? ☐ Yes ☑ No
4. Give dates of any trial period or fixed trial date ____________
5. Level of judge District Judge
6. Parties to be served Defendants

In the	Backwater County Court
Claim no.	[As Claim Form]
Warrant no. (If applicable)	
Claimant (including ref.)	Tumbledown Management Limited
Defendant(s) (including ref.)	Holly Andrews (1) and Ivy Andrews (2)
Date	[Date]

Note You must complete Parts A **and** B, **and** Part C if applicable. Send any relevant fee and the completed application to the court with any draft order, witness statement or other evidence; and sufficient copies for service on each respondent.

Part A

1. Enter your full name, or name of solicitor

~~I~~ (We)[1] Tumbledown Management Limited ~~(on behalf of)~~(the claimant)~~(the defendant)~~

2. State clearly what order you are seeking and if possible attach a draft

intend to apply for an order ~~(a draft of which is attached)~~ that[2]

for summary judgment under CPR, Part 24 and costs

because[3]

3. Briefly set out why you are seeking the order. Include the material facts on which you rely, identifying any rule or statutory provision

the Defendants have no real prospect of successfully defending the claim and we know of no other reason why disposal of the claim should await trial

Part B

~~I~~ (We) wish to rely on: *tick one box*

the attached (witness statement)(affidavit) ☐ my statement of case ☐

evidence in Part C in support of my application ☑

4. If you are not already a party to the proceedings, you must provide an address for service of documents

Signed ____________ (Applicant)~~('s Solicitor)('s litigation friend)~~

Position or office held Director (if signing on behalf of firm or company)

Address to which documents about this claim should be sent (including reference if appropriate)[4]

[Address details]	if applicable	
	fax no.	
	DX no.	
Tel. no. Postcode	e-mail	

The court office at

is open from 10am to 4pm Monday to Friday. When corresponding with the court please address forms or letters to the Court Manager and quote the claim number.

N244 Application Notice (4.00) *Printed on behalf of The Court Service*

Part C

Claim No. [As Claim Form]

~~I~~ (We) wish to rely on the following evidence in support of this application:

1. The Defence relies solely on the allegation that the service charges which make up the claim are unreasonable and excessive but the Defendants provide no particulars or evidence in support of this allegation.

2. In paragraph 7 of the Defence the Defendants say they will abide by any determinations made by the court or the LVT but they do not seek any such determinations.

3. In any event, the service charges which are the subject of this claim were determined by the LVT in other proceedings under case number [insert] on [insert date]. A copy of the LVT's decision is attached to this application.

4. In the circumstances the Defendants have no real prospect of succeeding on their Defence and we know of no other reason why disposal of the claim should await trial. We therefore ask for summary judgment in favour of the Claimant.

Statement of Truth

~~*(I believe)~~ *(The applicant believes) that the facts stated in Part C are true

**delete as appropriate*

Signed

(Applicant)~~('s Solicitor)('s litigation friend)~~

Position or office held Director

(if signing on behalf of firm or company)

Date [Date]

Notes on completing this form

As with other prescribed court forms, many of the deletions and additions will need to be made after the form has been printed, so it is important to remember to make them and to take copies of the final versions. The official notes attached to the form are on the last page; nonetheless it is advisable to read them through first. The notes below are supplementary and are not intended as a substitute for the official guidance notes.

An application for summary judgment is of the type which will almost always be dealt with at a hearing since, if it succeeds, it will be the final hearing in the case. Fifteen minutes is a minimal time estimate for a straightforward point. Thirty minutes, an hour, or even longer may be more appropriate. Some thought needs to be given to the complexity of any legal argument or the volume of evidence. In a small claim especially however, a longer time estimate is likely to be counter-productive as there is little point in the court entertaining an application for summary judgment which is expected to take longer than a full hearing.

An application of this sort will almost invariably be heard by the district judge.

The wording of the application in part A is paraphrased from the terms of PD 24 paragraph 2(3) of the CPR.

The application and the statement of truth at the end of part C should be signed by a competent person — usually the party or (if the party is a company) an officer of the company. It should not be the managing agent. Previous points concerning the seriousness of a statement of truth apply here equally.

Witness Statement

Made on behalf of the Claimant
Witness: E. Reddy
1st statement
Exhibits: ER1 to ER5
Made: *Date*

IN THE BACKWATER COUNTY COURT

Claim No. [*As Claim Form*]

BETWEEN

TUMBLEDOWN MANAGEMENT LTD

Claimant

AND

HOLLY ANDREWS (1) AND IVY ANDREWS (2)

WITNESS STATEMENT OF ETHEL REDDY

---

1. I am Ethel Reddy of 69 North Street, Backwater, Nossex. I am a partner in the firm of Reddy & Willing, managing agents for the Claimant. I have full knowledge of the facts of this case and I am duly authorised to make this statement on behalf of the Claimant. Insofar as the contents of this statement are within my personal knowledge they are true, otherwise they are true to the best of my knowledge, information and belief.

Remaining numbered paragraphs: insert the witness's factual evidence.

STATEMENT OF TRUTH

I believe the facts stated in this witness statement are true.

Signed:
Position:

Dated:

Notes on preparation of witness statements

The formalities required in small claims tend to be fewer and less rigidly observed; nonetheless professionally represented parties will have higher standards expected from them.

Statements must be on one A4 side only, with a wide margin (35mm) for references to exhibit numbers. Each page should be numbered and securely fastened to each other. At the top right-hand corner of the first page the following should appear:

- the party on whose behalf the statement is filed
- the initials and surname of the witness
- the number of the statement by that witness (1st, 2nd and so on)
- the initials and numbers of the exhibits attached (if any)
- the date of signature of the statement.

The statement must be headed by the title of the action (claim number, names of parties and so forth).

The main text should be in the witness's own words and in the first person. The opening paragraph should introduce and identify the witness, showing the basis of her interest in the proceedings and knowledge of the facts to follow. Where the witness refers to an exhibit, the reference should commence: "There is now shown to me marked ER1 (*or whatever it may be*) ..." followed by a description of the exhibited document.

As with other statements, the statement of truth must be taken seriously. A false statement may be punished as contempt of court.

Exhibits also have formalities which are similar to those for exhibits to an affidavit. They should have a frontsheet carrying the same headings as the statement, then wording to the effect of: "This is the exhibit marked ER1 referred to in the statement of Ethel Reddy dated [*date of statement*]".

Methods of enforcing a county court judgment

Warrant of execution

Despite its colourful name (and to the disappointment of many judgment creditors) a warrant of execution is merely the means by which a judgment debtor's goods may be seized by the county court bailiff and sold off to satisfy the debt — and it rarely gets as far as that. Indeed, many creditors regard this as a too frequently unsuccessful method of enforcement.

The judgment creditor requests a warrant by the prescribed county court form N323 which gives basic details of the outstanding debt. The form includes a space for completion by the creditor to alert the bailiff to any relevant information, which is used too sparingly.

When the bailiff visits he is subject to a number of restrictions:

- he cannot gain access to the debtor's premises by force
- goods can be seized only to the equivalent value of the debt
- "goods" excludes tools, vehicles or other equipment necessary for the debtor's trade and clothes, furniture or other domestic necessaries of the debtor or his family
- only goods belonging personally to the debtor may be seized. All

parties (including the bailiff and the judge) may become embroiled in lengthy "interpleader" proceedings if there is doubt about ownership.

Although goods may have been "seized" by the bailiff, what normally happens is that the bailiff takes what is known as "walking possession". This means that the goods stay where they are with a responsible person undertaking not to remove them until the debt has been discharged (including the costs of the warrant procedure) or they are taken away for sale. In such a case, the bailiff is entitled to use force if necessary to enter the premises to regain possession of the goods.

In the unlikely event things progress to that point, the goods will be sold at auction and the debtor receives any surplus after the debt and costs are discharged.

Most frequently, either the bailiff's visit is enough to prompt an offer of payment or he simply cannot gain peaceful entry to the debtor's premises. Indeed, many bailiffs' attempts to execute default judgments lead to debtors' applications to set aside the judgments.

Third party debt order

A third party debt order is the modern version of garnishee proceedings. The new title is almost self-explanatory: it is an order by the court to a third party who owes money to the debtor or holds money for him to pay the judgment creditor instead. Since the third party has no axe to grind with the creditor (normally), it is an effective way of obtaining payment but the path there is not straightforward.

The first step is an application on county court form N349 by which the creditor verifies (backed up by a statement of truth) the information required by Part 72 of the CPR, including:

- the details of the judgment debt, the creditor and debtor
- the amount outstanding
- the identity of the third party and, if it is a banking institution, details to help trace the account (the form makes it obvious that this is usually a bank or building society believed to hold funds for the debtor)
- details of any other person with an interest in the same funds
- the source of the creditor's knowledge or information (which may be following the performance of an order to obtain information from judgment debtors — see below).

As can be seen, the requirements for this application are taxing, but this is no doubt because the application can be made without notice to either the debtor or the third party (to avoid the risk of the funds being dissipated or redirected).

The application goes to the judge for a paper consideration (although obviously speculative applications are likely to be weeded out first). If the judge approves it, he or she will make an interim order which (a) directs the third party not to reduce the fund to less than the judgment debt (including the costs of the application), and (b) fixes a hearing to consider making the order final.

The interim order is served on both the third party and the debtor. If the third party is a bank, it must search its records and disclose to the court and the creditor any account numbers for the debtor, whether they are in credit, and whether they hold sufficient funds to discharge the order. Other third parties must disclose whether nothing is owed to the debtor or not enough to satisfy the order.

There are a few circumstances under which the court may refuse to grant a final order even if the funds appear to be present (for example, on the intervention of another person with an interest in the funds) but generally, if it has been established that sufficient monies exist to discharge the debt, the order will be made final and the third party required to hand over to the creditor enough to clear the judgment debt and the costs allowed on the application.

Charging order

The idea of a charging order was often sniffed at by landlords in the days when forfeiture remained a real possibility, because forfeiture superseded any mortgages or legal charges on the lease. Why should a landlord go to all the trouble of taking proceedings only to end up with something inferior to what he already possessed? As the availability of forfeiture has diminished, so has the attractiveness of charging orders increased.

The purpose of a charging order is to secure a judgment debt upon the debtor's interest in a property. It does not of itself realise any cash, but the charge is registrable as a caution at HM Land Registry so it will have to be cleared before the debtor can deal in his property; alternatively, the creditor may be able to obtain an order that the property be sold to release the funds.

An application is made on county court form N379, the essential elements of which are:

- details of the judgment and the outstanding debt
- the address of the property to be charged together with its title number (an office copy of the entries at HM Land Registry should be attached)
- a statement of the debtor's interest in the property and (if this is not obvious from the office copy entries) how the judgment creditor knows of the debtor's status
- details of other creditors, insofar as they are known to the applicant
- details of other persons with an interest in the property who need to be served with the application (for example a spouse or mortgage lender)
- a statement of truth on the part of the person signing the application.

The application (which is made without notice to the debtor initially) goes before the district judge for a paper consideration with a view to making an interim order. The interim order (which used to be known as a charging order nisi) includes a hearing date for a final order. This interim order can — and should — be registered as a caution against dealings in the property at HM Land Registry.

The interim order and copies of the application and any supporting paperwork must be served upon the debtor and other persons specified by the court (who will include the other interested parties identified in the application). The court may order the judgment creditor to effect service and certify that it has been done. Service of the order effectively prevents any dealing in the property until the final outcome of the application is known.

In considering whether to make the order final, the court must take into account the personal circumstances of the debtor and whether any other creditor would be prejudiced unduly by the order. The court may also refuse to make the order final if there are cogent objections from interested parties, which might include doubts over title or evidence that there is insufficient equity in the property. Indeed, there is little point in pursuing a charging order application if it is obvious that there is insufficient equity and that there is unlikely to be for the foreseeable future.

A final charging order (previously a charging order absolute) should also be registered at HM Land Registry, and it will rank behind any previously registered mortgages and legal charges.

A charging order may be followed up by an application for an order for sale under Part 73 of the CPR. This application requires new, separate proceedings commenced with a part 8 claim form (the equivalent of an old originating application) rather than being part and parcel of the proceedings which comprised the original judgment and the charging order. The evidence in support of the application must detail the charging order plus up to date verifications of the same general issues which went to support the charging order application in the first place. There must also be a valuation of the property or at least a realistic estimate of the price obtainable on sale. The claimant will frequently apply for an order that those in occupation vacate the premises for the purposes of a sale with vacant possession.

Where the judgment debtor is a joint owner of the property, the situation is inevitably more complicated. Applications for orders for sale in such circumstances are made under the Trusts of Land and Appointment of Trustees Act 1996. Legal advice and representation is strongly recommended in these cases.

Attachment of earnings order

A judgment creditor may apply on county court form N337 for an order that the judgment debtor's employer makes deductions from the debtor's remuneration over a period of time (the amounts and rates to be set by the court) until the debt is satisfied. Naturally, the application can only be made if the creditor can establish that the debtor is employed and by whom. Moreover, the entitlement to an attachment of earnings order only arises if and when the debtor has failed to make a payment under an instalment order (for example, an order made following an offer to pay within an admission).

The application is served upon the debtor by the court, together with a questionnaire concerning the debtor's means. The completed questionnaire is copied to the creditor and it is considered by a court clerk, who may make the order if all the information is to hand. If either party objects or if the clerk feels he cannot make the order, the matter is listed for hearing before the district judge.

Presuming that the facts of the application are correct, the order will be made on the basis of fixed formulae for ascertaining the payment rate. The order will not reduce the debtor's income below the "protected earnings rate" (that is the minimum amount considered necessary to maintain the debtor and his family). The order is then served upon the employer, who faces possible contempt proceedings if he fails to comply.

While an attachment of earnings order is in place, the creditor needs the court's permission for a warrant of execution.

Order to obtain information from judgment debtors

As stated in Chapter 7, this method (formerly known as an "oral examination") is not strictly a means of enforcement, but it may supply vital information showing the most suitable approach and, in any event, the stigma attached may be sufficient to encourage payment. There is a prescribed form (county court form N316) which, apart from the obvious questions, asks the judgment creditor to consider a number of options, including:

- questions to be put to the debtor over and above the standard points in form EX140 — often a landlord or management company will have specific knowledge of a leaseholder's assets which might throw up a valuable line of enquiry, but this should be used only for the purpose of establishing a means of enforcing the debt
- on the same basis, the debtor may be ordered to produce financial documents in addition to those listed in form EX140: these may be personal or business accounts
- the examination will normally be conducted by a court clerk but, if special reasons can be shown, the judge may do it.

The order to attend must be served upon the judgment debtor personally, and it is usually the job of the judgment creditor to arrange this. The debtor may request travelling expenses from the creditor. The creditor is required to swear an affidavit certifying service and stating whether or not travelling expenses were requested and paid. The affidavit must also verify how much of the debt remains outstanding.

The examining officer will put the questions at the hearing. The creditor may ask supplementary questions or ask the court officer to put them. The examination is intended to be "of the severest kind". Information gleaned at the hearing can be used to substantiate any of the other forms of enforcement.

If the debtor fails to attend or otherwise fails to comply with the procedure, it is usual for the court to make a suspended committal order (an order for committal to prison for contempt of court) which effectively gives the debtor a last chance. In the event of further non-compliance, the debtor will be brought before the judge to show why he should not be sent to prison.

Appendix 5
Forfeiture for Breach of Repairing Covenant

Section 146 notice

NOTICE UNDER SECTION 146 OF THE LAW OF PROPERTY ACT 1925

TO Louis and Susan Hoodoo the tenants of the premises known as Flat 12A Tumbledown Mansions, Dystopia Road, Backwater, Nossex comprised in a Lease dated 24 June 1985 and made between Reston Laurels plc as Lessor (1) and Tumbledown Management Ltd as Manager (2) and Jonah Jeremiah Job as Lessee (3)

Whereas the interest of Jonah Jeremiah Job has been assigned to Louis and Susan Hoodoo

We Reston Laurels plc of Freeholder House, West Street, Backwater, Nossex HEREBY GIVE YOU NOTICE as follows:

1. Clause [*insert clause number*] of the said Lease contains a covenant by you or your predecessor in title in the following terms:
   [*insert wording of repairing clause — repeat if more than one covenant is to be cited*]
2. The said covenant has been broken and the particular breach complained of is that you have failed to keep the premises in

repair [*or something similar, following the wording of the covenant as closely as possible*] as appears from the attached Schedule of Dilapidations and Wants of Repair prepared by Richard Fellows FRICS and dated [*insert date of schedule*]

3. We require you to remedy the said breach or breaches insofar as the same may be capable of remedy
4. We further require you to pay compensation in money for the breach or breaches and also to pay pursuant to clause [*insert number of leaseholder's costs covenant*] of the Lease [*insert relevant particulars of the leaseholder's obligations to pay costs as described in the lease*]
5. If you fail to comply with this Notice within a reasonable time we will exercise our right of re-entry under clause [*insert number of forfeiture clause*] of the Lease and claim damages for the said breach or breaches of covenant
6. You have the right to serve a Counter-Notice claiming the benefit of the Leasehold Property (Repairs) Act 1938. Such Counter-Notice must be in writing and must be to the effect that you claim the benefits of that Act and must be served within 28 days of the service upon you of this Notice. It may be served by leaving it at the address of the Landlord's Agents set out below or by sending it there by post in a registered letter or a letter sent by the recorded delivery service and provided it is not returned through the post undelivered if you use such a letter your Counter-Notice will be deemed to be served at the time when the letter would in the ordinary course be delivered

DATED this [*number*] day of [*month and year*]

Signed ______________________________

Swift & Sharp of 45 East Street, Backwater, Nossex
Solicitors and Agents for Reston Laurels plc

Notes on completing a section 146 notice

There is no prescribed form for a section 146 notice, but the section itself sets out the fundamental requirements, and they have been fleshed out since by case law. The example above has been tried and tested.

It is best to avoid putting in too many precise figures or time limits at this stage; they can be negotiated later. If the breaches complained of are or include non-payment of monies then those figures must be included, as section 146 requires that the notice specifies the breaches.

This general form of notice may be used for all breaches of covenant except non-payment of rent, save that paragraph 6 is expressly inserted in repairs cases.

What is a "reasonable time" (paragraph 5)? There is no uniform answer. It has to be a period within which the recipient has a reasonable opportunity to remedy his breach. For payment of monies, this may be 14 or 21 days. For irremediable breaches a very short time will suffice. Specifically in repairs cases, the leaseholder has 28 days within which to serve a counter-notice, and the landlord may take no further steps within that time.

Appendix 6
Forfeiture for Other Breaches

Claim for a declaration

On the following pages a sample Claim Form is reproduced. This form can be downloaded from *www.hmcourts-service.gov.uk*

Claim Form
(CPR Part 8)

In the Backwater County Court	
Claim No.	

SEAL

Claimant
Reston Laurels Plc

Defendant(s)
Mr A Wohl

Does your claim include any issues under the Human Rights Act 1998? ☐ Yes ☑ No

Details of claim *(see also overleaf)*

The Claimant seeks a Declaration under Section 168 of the Commonhold and Leashold Reform Act 2002 that the Defendant is in breach of the covenants in his lease so that the Claimant is entitled to commence forfeiture proceedings against the Defendant

Defendant's name and address

Mr A Wohl
Flat 5, Tumbledown Mansions
Dystopia Road
Backwater
Nossex

	£
Court fee	150.00
Solicitor's costs	To be assessed
Issue date	

The court office at

is open between 10 am and 4 pm Monday to Friday. When corresponding with the court, please address forms or letters to the Court Manager and quote the case number.

N208 Claim form (CPR Part 8) (10.00) *Printed on behalf of The Court Service*

Claim No.	

Details of claim *(continued)*

1. Part 8 of the Civil Procedure Rules 1998 applies.

2. The Claimant seeks a Declaration on a question which is unlikely to involve a substantial dispute of fact. The Claimant wants the Court to decide that:-

a) The Defendant is in breach of covenant under the terms of the Lease dated 24th June 1985 for Flat 5, Tumbledown Mansions, Dystopia Road, Backwater, Nossex and made between Reston Laurels Plc (1) and Tumbledown Management Limited (2) and Romeo and Juliet Montagu (3)

b) The Claimant is entitled to commence forfeiture proceedings for the said breach of covenant

c) The Defendant should pay the Claimant's costs assessed summarily (and in accordance with the terms of the Lease described at a) above)

and to make a Declaration (in accordance with Section 168 of the Commonhold and Leasehold Reform Act 2002) in the terms of 2a) and b) above and to further order that the Defendant should pay the Claimant's costs and interest.

3. For further details of the claim, see attached Particulars of Claim.

Statement of Truth

*~~(I believe)~~(The Claimant believes) that the facts stated in these particulars of claim are true.

* I am duly authorised by the claimant to sign this statement

Full name Raymond Zoroaster Sharp

Name of claimant's solicitor's firm Swift & Sharp

signed ______________________ position or office held Partner

*~~(Claimant)(Litigation friend)~~(Claimant's solicitor) (if signing on behalf of firm or company)

delete as appropriate

Messrs Swift & Sharp
45 East Street
Backwater
Nossex

Ref: RZS

Claimant's or claimant's solicitor's address to which documents should be sent if different from overleaf. If you are prepared to accept service by DX, fax or e-mail, please add details.

Notes to completing N208 Claim Form

As with all prescribed forms, they are subject to change. Practitioners are advised to carry out regular checks to ensure that their stocks are up to date. Electronic versions of prescribed forms will generally prompt insertions at the appropriate points; however, some points need to be deleted and others need to be typed in somehow, depending upon the available technology. Many of the deletions and additions will need to be made after the form has been printed, so it is important to remember to make these and to take copies of the final versions.

Form N208 is relatively straightforward. The word "declaration" should be flagged up under the heading "details of claim" on the first page so that the relevant procedures are applied from the start. Similarly, to assist the court with identifying the type of case, it should refer to the fact that it is a part 8 claim and that it the application is made in the context of the Commonhold and Leasehold Reform Act 2002.

The points made previously in relation to statements of truth apply equally to this form.

The claim form will be filed with attached particulars of claim as set out below. This is a word processor document created by the claimant's solicitors rather than a prescribed form.

IN THE BACKWATER COUNTY COURT

CLAIM NUMBER:

BETWEEN

RESTON LAURELS PLC

Claimant

And

MR A WOHL

Defendant

PARTICULARS OF CLAIM

---

1. Under a Lease ("the Lease") dated 24th June 1985 the Claimant let the property known as Flat 5, Tumbledown Mansions, Dystopia Road, Backwater, Nossex ("the Property") to Romeo and Juliet Montagu.

2. The Lease was for a fixed term of 99 years from 24th June 1985. A copy of the Lease is annexed at Attachment A and it should be referred to for its full terms and effect.
3. The Defendant was registered with the leasehold title to the property on [*insert date*] following a purchase which appears to have taken place on [*insert date*]. Office copy entries of the Claimant's and Defendant's titles are annexed at Attachment B.
4. Under the Lease the Defendant covenanted at clause [*insert number*] as follows:
   "Not at any time during the term to assign underlet or otherwise part with possession of part only of the demised premises as opposed to the whole".
5. Further under clause [*insert number*] of the Lease the Defendant covenanted *[insert details of costs covenant (assuming it is relevant)*].
6. There is a forfeiture clause at clause [*insert number*] of the Lease.
7. The Defendant is in breach of the covenant contained in clause [*insert number of clause quoted in 4 above*] of the Lease.
8. The Property consists of a flat and garage. On or about [*insert approximate date*] the Defendant let the flat to Mr and Mrs T Dance. Subsequently on or about [*insert approximate date*] and in breach of covenant the Defendant let the garage separately to Mr Christopher Carr. Furthermore Mr Carr is using the garage in such a way as to obstruct the rights of way of other leaseholders at Tumbledown Mansions. The Claimant relies upon the written evidence of [*name of witness*] served with this claim.
9. The name and address of the person to be served is Mr A Wohl of [*insert appropriate address details*].

Dated ____________________

Signed ____________________

Messrs Swift & Sharp
45 East Street
Backwater
Nossex

Solicitors for the Claimant

Notes for completing the particulars of claim

Paragraphs 1–3. It is essential to include the lease and sufficient details to show how the parties acquired their respective interests in the property.

Paragraph 4 can be repeated if there is more than one covenant being relied upon, or alternatively they can all be set out as sub-paragraphs within paragraph 4.

Paragraph 5 is useful to set out the contractual claim for costs but, if there is not a relevant clause in the lease, this paragraph should be omitted.

Paragraph 7 can be pluralised if necessary to list the total number of covenants breached by the defendant.

Paragraph 8. Evidence should be supplied from a relevant witness (here, the managing agent or a director of TML) to support the particulars of the alleged breach or breaches and to show the efforts that have been made to resolve the matter before issuing the proceedings. None the less, the particulars of claim should include at least a brief description of what constitutes the breach.

Indeed, particulars of claim are not strictly necessary in a part 8 claim. The alternative way to deal with it would be combine the particulars and the supporting evidence in one witness statement. Different practitioners will have their own approaches and different courts may have their own requirements. Trial and error will disclose the most suitable strategy to meet local conditions over a period of time.

Section 146 notice

NOTICE UNDER SECTION 146 OF THE LAW OF PROPERTY ACT 1925

TO Mr A Wohl the tenant of the premises known as Flat 5 Tumbledown Mansions, Dystopia Road, Backwater, Nossex comprised in a Lease dated 24 June 1985 and made between Reston Laurels plc as Lessor (1) and Tumbledown Management Ltd as Manager (2) and Romeo and Juliet Montagu as Lessee (3)

Whereas the interest of Romeo and Juliet Montagu has been assigned to Mr A Wohl

We Reston Laurels plc of Freeholder House, West Street, Backwater, Nossex HEREBY GIVE YOU NOTICE as follows:

1. Clause [*insert clause number*] of the said Lease contains a covenant by you or your predecessor in title in the following terms:
   "Not at any time during the term to assign underlet or otherwise part with possession of part only of the demised premises as opposed to the whole".
2. The said covenant has been broken and the particular breach complained of is that on or about [*insert approximate date*] you have underlet or otherwise parted with possession of part only of the demised premises (namely the garage)
3. We require you to remedy the said breach or breaches insofar as the same may be capable of remedy
4. We further require you to pay compensation in money for the breach or breaches and also to pay pursuant to clause [*insert number of leaseholder's costs covenant*] of the Lease [*insert relevant particulars of the leaseholder's obligations to pay costs as described in the lease*]
5. If you fail to comply with this notice within a reasonable time we will exercise our right of re-entry under clause [*insert number of forfeiture clause*] of the Lease and claim damages for the said breach or breaches of covenant

DATED this [*number*] day of [*month and year*]

Signed ____________________

Swift & Sharp of 45 East Street, Backwater, Nossex
Solicitors and Agents for Reston Laurels Plc

(*Note* See the notes on completing a section 146 notice in Appendix 5.)

Glossary of Terms

Administration charge	A charge payable under a lease by a leaseholder for approvals, provision of information or as a result of a breach of his obligations (see Chapter 13)
AIRPM	Associate of the Institute of Residential Property Management
All ER	*All England Law reports*
ASBI	Anti-Social Behaviour Injunction
ASBO	Anti-Social Behaviour Order
CCS	Commonhold Community Statement (see Chapter 11)
Commonhold association	A private company limited by guarantee the object of which is to administer a Commonhold development
Common law	The body of law in England and Wales based on the historical development of judicial decisions and the legal doctrines and principles governing them
Counsel	A barrister
Covenant	A binding promise embodied in a contract
CPR	The Civil Procedure Rules, the system of rules and procedures governing the practice of the civil courts including the High Court and the county court
Demised premises	The flat, apartment or house and any associated land (including, for example, garages) which are let to the leaseholder
Disbursements	Out of pocket expenditure, usually by solicitors or agents, as distinct from fees or other profit-making charges
Distress	The seizing of goods belonging to a debtor as security or payment for the debt
EGLR	*Estates Gazette Law Reports*

Enfranchisement	(In a leasehold context) The acquisition of the free-hold interest by the leaseholders
Equity	The body of law based on the application of natural justice historically applied to modify the more rigorous effects of common law and statute
Forfeiture	(In a leasehold context) The early termination of a lease by the landlord
Freehold Ieversion	The freeholder's residual interest in leasehold property which comes to fruition at the end of the term of the lease
Garnishee	A third party who owes money to or holds money for a debtor and who may be called upon to pay a creditor instead up to the value of a judgment debt
Injunction	An order of the court preventing or restraining someone from doing something or, occasionally (as a mandatory injunction), an order requiring someone to perform a positive act
Interpleader	A court procedure to determine competing interests of claimants to property or goods, or someone who invokes such a procedure
IRPM	Institute of Residential Property Management
Landlord	The party contractually entitled to enforce the covenants of a leaseholder or tenant (which could be a freeholder, a resident-owned management company, a right to manage company or a right to enfranchise company)
LB	London borough
LBC	London Borough Council
Leaseholder	A tenant under a long lease
Lessee	A tenant under a long lease
Lessor	A landlord under a long lease
Licensed conveyancer	A lawyer specialising in conveyancing who is licensed by the Council for Licensed Conveyancers to offer conveyancing services direct to the general public
Limitation	A period imposed by statute after which any legal claims within the relevant category are barred
Litigant in person	A party to court proceedings who conducts his or her case without legal representation
Litigation	Legal proceedings
LVT	Leasehold Valuation Tribunal
Mesne profits	A charge for use and occupation for premises in lieu of rent when the landlord is being kept out of possession (for example, during the period between an order for possession and the execution of the order)

Mortgagee in possession	A mortgage lender which has exercised its right under a mortgage deed to take possession of the mortgaged property (usually with the benefit of a court order after the borrower has defaulted on re-payment of the loan)
PD	A Practice Direction under the Civil Procedure Rules (see "CPR")
PTR	Pre-trial review, a preliminary hearing of the court or tribunal, usually for the purpose of giving procedural directions
Privity	A direct interest in a contract or its subject-matter, close enough to establish legal rights and liabilities
Quarter day	(In the leasehold context) One of the four calendar dates historically or traditionally fixed for payments of rent (25 March, 24 June, 29 September and 25 December)
QC	Queen's Counsel; a senior barrister, appointed to serve nominally as counsel to the Crown
Quiet enjoyment	The entitlement (of a leaseholder) to use and enjoy demised premises for the term of a lease without interference or interruption
RMC	A residents' management company; a company owned by leaseholders for the purpose of managing their block of flats or estate (often also owning the freehold of the property)
RTE	Right to Enfranchise (under the Commonhold and Leasehold Reform Act 2002)
RTM	Right to manage (under the Commonhold and Leasehold Reform Act 2002)
Re-entry	(In the leasehold context) The landlord's power to resume possession of leasehold property at the end of the term of the lease (usually in the event of forfeiture)
Relief from forfeiture	An order of the court or an agreement by the landlord relieving a leaseholder from the effects of forfeiture and reinstating the lease
Right of audience	Entitlement or permission to address the court or tribunal
Section 146 notice	A notice preliminary to forfeiture given under section 146 of the Law of Property Act 1925
Service charge	A charge payable under a lease for services, repairs, maintenance, improvements or insurance or the landlord's costs of management (see Chapter 6)
Silk	A colloquial term for a QC
Social landlord	A provider of social housing (generally a local housing authority, housing association or housing action trust)
Statute	Legislation; an Act of Parliament

Tenant	One who has the use and enjoyment of land, a building or part of a building, or other property owned by another for a fixed or periodic term in return for the payment of rent (and any other terms as agreed between the parties)
Unless orders	Orders of the court imposing a sanction or penalty in the event of a party's failure to comply with other terms of the order
Waiver	(In the leasehold context) An act or omission by which a party dispenses with the use of rights and remedies and is thus barred from exercising them
Wasted costs	Costs incurred by a party for a purpose which was circumvented by the act or omission of another

Useful Addresses

AIMS: Advice, Information and Mediation Service for Retirement and Sheltered Housing
Age Concern England
Astral House
1268 London Road
London SW16 4ER
Tel: 020 8765 7465
www.ageconcern.org.uk or *email aims@ace.org.uk*

ARMA: Association of Residential Managing Agents
178 Battersea Park Road
London SW11 4ND
Tel: 020 7978 2607
www.arma.org.uk

ARHM: Association of Retirement Housing Managers
3rd Floor, 89 Albert Embankment
London SE1 7TP
Tel: 020 7820 1839
www.arhm.org

FPRA: Federation of Private Residents' Associations
59 Mile End Road
Colchester
Essex CO4 5BU
Tel: 0871 200 3324
www.fpra.org.uk

HM Courts Service headquarters
5th Floor, Clive House
Petty France
London SW1H 9HD
Tel: 020 7189 2000 or 0845 456 8770
www.hmcourts-service.gov.uk

IRPM: Institute of Residential Property Management
178 Battersea Park Road
London SW11 4ND
Tel: 020 7622 5092
www.irpm.org.uk

LEASE: Leasehold Advisory Service
70–74 City Road
London EC1Y 2BJ
Tel: 020 7490 9580 or 0845 345 1993
www.lease-advice.org

Leasehold Valuation Tribunal
Residential Property Tribunal Service
10 Alfred Place
London WC1E 7LR
Tel: 020 7446 7700
www.rpts.gov.uk

Office of the Deputy Prime Minister
Eland House, Bressenden Place
London SW1E 5DU
Tel: 020 7944 4400
www.odpm.gov.uk

Residential Property Tribunal Service
10 Alfred Place
London WC1E 7LR
Tel: 020 7446 7700
www.rpts.gov.uk

RICS: Royal Institution of Chartered Surveyors
12 Great George Street
Parliament Square
London SW1P 3AD
Tel: 0870 333 1600
www.rics.org

Index